Transportation Planning, Program and Investment Decision Making 2018

Transportation Research Record: Journal of the Transportation Research Board Volume 2672, Issue 51

ISBN: 978-1-5108-8256-0

Printed from e-media with permission by:

Curran Associates, Inc.
57 Morehouse Lane
Red Hook, NY 12571

Some format issues inherent in the e-media version may also appear in this print version.

Published by Sage Publications USA

ISBN (Print) 978-1-5108-8256-0
ISBN 2018 Printed Set (All Issues) 978-1-5108-7735-1

Printed by Curran Associates, Inc. (2019)

Additional copies of this publication are available from:

Curran Associates, Inc.
57 Morehouse Lane
Red Hook, NY 12571 USA
Phone: 845-758-0400
Fax: 845-758-2633
Email: curran@proceedings.com
Web: www.proceedings.com

Volume 2672 Number 51 December 2018

TRR
JOURNAL OF THE TRANSPORTATION RESEARCH BOARD

Contents

Transportation Planning, Program and Investment Decision Making

Articles

Selecting and Scheduling Link and Intersection Improvements in Urban Networks 1
Uros Jovanovic, Elham Shayanfar, and Paul M. Schonfeld

Transportation Planning for Connected Autonomous Vehicles: How It All Fits Together 12
Bobby J. Cottam

A Framework for Identifying Recurring Bottlenecks in the State Planning Process in Colorado 20
Oana Deselnicu and Jason Wallis

Is Performance Measurement Improving Planning Practice? The Case of Location Affordability in Long-Range Transportation Plans 31
Ann M. Hartell

An Integrated, Validated, and Applied Activity-Based Dynamic Traffic Assignment Model for the Baltimore-Washington Region 45
Lei Zhang, Di Yang, Sepehr Ghader, Carlos Carrion, Chenfeng Xiong, Thomas F. Rossi, Martin Milkovits, Subrat Mahapatra, and Charles Barber

Are Small- and Medium-Sized Community Paratransit Riders Ready to Adopt Real-Time Information (RTI) Technology? 56
Jeffrey J. LaMondia, Travis Gajkowski, and Veronica Ramirez

Strategically Aligning Capital Improvement Prioritization to Performance Goals 68
Elliot Sperling and Catherine Ross

Profile Monitoring for Examining Impact of Hurricane Katrina and Rita on Highway Construction Cost 79
Minsoo Baek and Baabak Ashuri

Illustrating the Monetary Impact of Transportation Projects/Alternatives using the Values of Travel Time and Travel Time Reliability 88
Venkata R. Duddu, Srinivas S. Pulugurtha, and Praveena Penmetsa

Programming Flexible Congestion Mitigation and Air Quality Program Funds: Best Practices for State DOTs 99
Gregory Rowangould, Razieh Nadafianshahamabadi, and Amir Poorfakhraei

Transportation Officials, Washington, D.C.

K. Jane Williams, *Acting Administrator, Federal Transit Administration, U.S. Department of Transportation*

Paul Zukunft *(Admiral, U.S. Coast Guard), Commandant, U.S. Coast Guard, Washington, D.C.*

---

* Membership as of February 2019.

TRANSPORTATION RESEARCH BOARD

CORE SPONSORS

Transportation Departments of the 50 States and the District of Columbia

Federal Government

- U.S. Department of Transportation
  - Federal Aviation Administration (FAA)
  - Federal Highway Administration
  - Federal Motor Carrier Safety Administration (FMCSA)
  - Federal Railroad Administration (FRA)
  - Federal Transit Administration
  - National Highway Traffic Safety Administration (NHTSA)
  - Office of the Assistant Secretary of Transportation for Research and Technology (OST-R)

- U.S. Air Force Civil Engineer Center (AFCEC)
- U.S. Coast Guard (USCG)
- U.S. Department of Energy
- U.S. Department of the Interior (DOI)
- U.S. Environmental Protection Agency (EPA)

Nongovernmental Organizations

- American Public Transportation Association (APTA)
- Association of American Railroads (AAR)
- California Air Resources Board

The National Academies of
SCIENCES · ENGINEERING · MEDICINE

The **National Academy of Sciences** was established in 1863 by an Act of Congress, signed by President Lincoln, as a private, nongovernmental institution to advise the nation on issues related to science and technology. Members are elected by their peers for outstanding contributions to research. Dr. Marcia McNutt is president.

The **National Academy of Engineering** was established in 1964 under the charter of the National Academy of Sciences to bring the practices of engineering to advising the nation. Members are elected by their peers for extraordinary contributions to engineering. Dr. C. D. Mote, Jr., is president.

The **National Academy of Medicine** (formerly the Institute of Medicine) was established in 1970 under the charter of the National Academy of Sciences to advise the nation on medical and health issues. Members are elected by their peers for distinguished contributions to medicine and health. Dr. Victor J. Dzau is president.

The three Academies work together as the National Academies of Sciences, Engineering, and Medicine to provide independent, objective analysis and advice to the nation and conduct other activities to solve complex problems and inform public policy decisions. The Academies also encourage education and research, recognize outstanding contributions to knowledge, and increase public understanding in matters of science, engineering, and medicine.

Learn more about the National Academies of Sciences, Engineering, and Medicine at **www.national-academies.org.**

The **Transportation Research Board** is one of seven major programs of the National Academies of Sciences, Engineering, and Medicine. The mission of the Transportation Research Board is to increase the benefits that transportation contributes to society by providing leadership in transportation innovation and progress through research and information exchange, conducted within a setting that is objective, interdisciplinary, and multimodal. The Board's varied activities annually engage about 7,000 engineers, scientists, and other transportation researchers and practitioners from the public and private sectors and academia, all of whom contribute their expertise in the public interest. The program is supported by state transportation departments, federal agencies including the component administrations of the U.S. Department of Transportation, and other organizations and individuals interested in the development of transportation.

Learn more about the Transportation Research Board at **www.TRB.org**.

Article

TRR
JOURNAL OF THE TRANSPORTATION RESEARCH BOARD

Transportation Research Record
2018, Vol. 2672(51) 1–11

DOI: 10.1177/0361198118758681
journals.sagepub.com/home/trr
SAGE

Selecting and Scheduling Link and Intersection Improvements in Urban Networks

Uros Jovanovic[1], Elham Shayanfar[1], and Paul M. Schonfeld[1]

Abstract

Deciding which projects, alternatives or investments to implement is a complex and important problem not only in transportation engineering, but in management, operations research and economics. Projects are interrelated if their benefits or costs depend on which other projects are implemented. Furthermore, in the network development problem analyzed here, the timing of projects also affects the benefits and costs of other projects. This paper presents a method for optimizing the selection and scheduling of interrelated improvements in road networks that explicitly considers intersections. The Frank Wolfe algorithm, which is modified here to consider intersections, is used for evaluating network improvements as well as for traffic assignment. Intersections are modelled with pseudo-links whose delays are estimated with Akcelik's generalized model. The objective is to minimize the present value of total costs (including user time) by determining which projects should be selected and when they should be completed. A genetic algorithm is used for optimizing the sequence and schedule of projects.

For decades transportation engineers have been dealing with the problem of evaluating, selecting and scheduling infrastructure projects. Considered alternatives can be classified as follows:

- Mutually exclusive: Only one alternative may be selected;
- Independent: The benefits and costs of alternatives are independent of which alternatives are selected or when those are implemented;
- Interdependent (interrelated).

Interrelated alternatives pervade transportation networks since improvements alter the flows, and hence benefits, on other network components. This paper aims to show how a traffic assignment model can be used to evaluate the objective function of an investment planning optimization problem for an urban road network, especially by showing how intersections can be included in the traffic assignment. A method is presented for evaluating, selecting and scheduling interdependent improvement alternatives in urban road networks, which extends Shayanfar et al. (*1*) by considering intersection improvements in addition to link widening alternatives. It is shown how a traffic assignment model can be effectively modified to consider intersection flows and delays by introducing pseudo-links. Adding pseudo-links for each of three movements (left, through and right) at each approach of a four-leg intersection, creates a total of 12 pseudo-links per intersection. Moreover, a traffic assignment model is shown to be effectively combined with a genetic algorithm for planning and prioritizing purposes while considering interrelations among candidate projects. The background section reviews some prior studies on intersection delay, selection and scheduling of project alternatives, and traffic assignment. The next two sections present the evaluation model and the genetic algorithm used for optimizing the project selection and schedule. A case study is presented on the Sioux Falls network and the results obtained with the modified traffic assignment model and genetic algorithm in optimizing the network development schedule. Conclusions and suggestions for extensions are presented in the last section.

Background

Intersections are crucial components in urban road networks since they affect traffic capacity and delay at least

[1]Department of Civil and Environmental Engineering, University of Maryland, College Park, MD

Corresponding Author:
Address correspondence to Paul M. Schonfeld: pschon@umd.edu

as much as road links. Typically, four-leg intersections allow up to 12 legal vehicular movements and 4 legal pedestrian crossing movements. Traffic signals assign right-of-way, and can significantly reduce the number of conflicts, thus regulating the traffic flow. One of the many disadvantages of traffic signals is the possibility of excessive delay which can congest the network, which, in turn, increases cost, pollution and driver anxiety. Early studies on delays at signalized intersections include Wardrop (*2*), who assumed that vehicles enter intersections with uniform arrivals, and Webster (*3*) who studied delays for vehicles at pretimed signals and optimized their settings.

Delay relates to the amount of excess travel time, fuel consumption, and the frustration and discomfort of drivers. Delay can also be used to compare the performance of an intersection under different demand, control and operating conditions. For intersections, delay can be calculated simply, as the difference in the departure time and the arrival time of a vehicle. Estimation of overflow delay is one of the major difficulties in developing delay models at signalized intersections. The difficulty is obtaining a simple and easily computable formula for overflow delay and has forced researchers and analysts to search for approximations and boundary values.

Numerous intersection delay models have been developed, including Webster's (*3*), Highway Capacity Manual (HCM) (*4*), Australian (variation of the Akcelik delay model (*5, 6*), and Canadian (*7*). The delay model used here is Akcelik's, because it gives delay values close to the HCM formula for $v/c < 1.0$, but with fewer assumptions about parameters. It is expressed in Equations 1 and 2 as

$$d = 0.5\frac{C(1-\lambda)^2}{(1-\lambda x)} + 900Tx^n\left[(x-1) + \sqrt{(x-1)^2 + \frac{m(x-x_0)}{cT}}\right] \quad (1)$$

and

$$x_0 = a + bsg \quad (2)$$

where

d = average overall delay (sec/veh),

C = cycle time (sec),

λ = fraction of the cycle which is effectively green for the phase under consideration,

x = v/c ratio,

T = flow period (h),

c = link capacity (veh/h),

m, n, a, b = calibration parameters, whose values are available for different delay models (e.g., Australian, Canadian, TRANSYT (*8*), and HCM) in Akcelik's paper (*5*), and

s^*g = capacity per cycle ($veh/cycle$).

Parameters n, m, a and b according to Akcelik's papers (*5, 6*) have the following values respectively: 0, 8, 0.5, and 0. Therefore, the two equations above become

$$d = 0.5\frac{C(1-\lambda)^2}{(1-\lambda x)} + 900T\left[(x-1) + \sqrt{(x-1)^2 + \frac{8(x-0.5)}{cT}}\right] \quad (3)$$

$$x_0 = 0.5$$

The overall delay d_I at an intersection can be calculated as

$$d_I = \frac{\sum d_A v_A}{\sum v_A} \quad (4)$$

where d_A is delay on approach A, and v_A is volume on approach A.

Heidemann (*9*) and Olszewski (*10*) used probability distribution function to estimate delay at signalized intersections. In their models, the probability distributions of delay were obtained from the probabilities of queue lengths.

Among many approaches used to tackle the problem of project selection and scheduling are integer programming, used by Weingartner (*11*) and by Cochran et al. (*12*), and dynamic programming, used by Weingartner (*11*) and by Nemhauser and Ulman (*13*). One notable study on interrelated projects is Weingartner's (*11*) which presents, among other problems, interdependent projects with budget constraints.

Mehrez et al. (*14*) use a multi-attribute function to specify the decision maker's preference with a zero-one budget model to solve the problem of selection of interrelated multi-objective long-range projects. The authors define a set of n indivisible projects contributing to m tangible and intangible attributes with L limited resources available for T periods. In addition, they use a utility function with m attributes and regard each project as a collection of subprojects, each one contributing to one of the attributes affected by the projects.

Evaluation Model

Traffic assignment can be formulated as the problem of finding the equilibrium flow pattern over a given transportation network, if its graph representation, the associated link performance function and an origin–destination (O-D) matrix is known. Assignment of traffic flows on network links is a result of equalizing transportation demand (O-D matrix) and transportation supply (link and node capacity, management actions). A reasonable assumption is that all travelers try to minimize their own travel time between their own origins and destinations. Other assumptions are that travel times increase with link flows, and all individuals behave identically. User equilibrium (stable condition) is achieved when no traveler can improve their travel time by changing route. Notable publications that dealt with traffic assignment include Florian (*15*), Sheffi (*16*), and Boyce and Ran (*17, 18*). However, none of these consider intersection characteristics and performance.

This paper applies the convex combination algorithm developed by Frank and Wolfe (*19*) to evaluate link and intersection expansion projects upon their implementation in the network. The Frank-Wolfe (FW) algorithm is an iterative algorithm used for solving a user equilibrium traffic assignment which is a nonlinear programming problem with convex objective function and linear constraints. Given t_a^0 (initial travel time for link a), the convex combination algorithm is as follows:

1. Set counter $n = 1$,
2. Initialization: Perform all or nothing assignment assuming t_a^{n-1}, which yields flows x_a^n
3. Update: Set link travel time (Bureau of Public Road (BPR) function) $t_a^n = t_a(x_a^n) = t_0\left(1 + 0.15\left(\frac{v}{c}\right)^4\right)$
4. Direction finding: Perform all-or-nothing assignment based on $\{t_a^n\}$, which will yield a set of (auxiliary) flows $\{y_a^n\}$
5. Line search: find $\propto_n$ that solves the following problem:

$$\max_{0 \le x \le 1} \sum_a \int_0^{x_a^n + \alpha(y_a^n - x_a^n)} t_a(\omega) d\omega \tag{5}$$

6. Move: Set $x_a^{n+1} = \alpha(y_a^n - x_a^n), \forall \alpha$
7. Convergence test: If a convergence criterion is met, stop. Otherwise, set $n = n + 1$ and return to step 2.

Problem Statement

The problem considered here is NP hard (*20*) with a non-convex objective function. The problem grows rapidly as the number of candidate projects increases, and can be classified as a combinatorial optimization problem. This type of problem involves finding values for discrete variables in such a way that the optimal solution is found with respect to the objective function. Many practical problems can be classified as combinatorial optimization problems such as the shortest path algorithm. Other examples are the optimal assignment of employees to tasks to be performed and the traveling salesman problem. Dorigo et al. (*21*) formulated a combinatorial optimization problem Π as a triple (S, f, Ω), where S is the set of candidate solutions (sequence of projects), f is the objective function (present value of total costs) which assigns an objective function value $f(s)$ to each candidate solution $s \in S$, and Ω is the set of constraints (budget constraint in our case). The solutions belonging to the set $\tilde{S} \subseteq S$ of candidate solutions that satisfy the constraints Ω are called feasible solutions. The goal, according to Dorigo et al. (*21*), is to find a globally optimal feasible solution s^* (optimal sequence of projects).

In this study, the present value of total cost during the analysis period is the objective function, subject to a budget constraint. The total cost consists of: (i) supplier cost, defined as the present value of all project costs, and (ii) user cost, defined as the delay multiplied by the value of time. Accordingly, the objective function can be formulated as

$$Z = \sum_{j=1}^{T} \frac{v}{(1+r)^j} \sum_{i=1}^{n_l} w_{ij} + \sum_{j=1}^{T} \frac{1}{(1+r)^j} \sum_{i=1}^{n_{pl}} c_i x_i(t) + \sum_{j=1}^{T} \frac{v}{(1+r)^j} \sum_{i=1}^{n_I} d_{ij} + \sum_{j=1}^{T} \frac{1}{(1+r)^j} \sum_{i=1}^{n_{pI}} C_i X_i(t) \tag{6}$$

where
w_{ij} = waiting time on link i in year j,
c_i = present value of the cost of link project i,
n_{pl} = number of link projects (link improvements),
n_l = total number of links,
n_I = total number of intersections,
n_{pI} = number of intersection improvement projects,
C_i = present value of the cost of intersection project i,
v = value of time, and
r = interest rate.

The cost of intersection project i can be written as

$$C_i = C_{c_i} + C_{p_i} = A_{I_i} \cdot 51.6 + A_i \cdot 20 \tag{7}$$

where
C_{c_i} = capital cost of improvement of intersection i ($\$/ft^2$),
C_{p_i} = cost of pavement maintenance of intersection i ($\$/ft^2$),
A_{I_i} = area of the land needed to improve intersection i (ft^2),
A_i = overall area of the intersection i (ft^2)

The objective function is bound by the following cumulative budget constraint (*22*) as

$$\sum_{i=1}^{n_p} c_i x_i(t) \le \int_0^t B(t) dt, 0 \le t \le T \tag{8}$$

$$\begin{cases} x_i(t) = 0 \text{ if } t < t_i \\ x_i(t) = 1 \text{ if } t > t_i \end{cases}$$

where t_i is the time when project i is finished, and $x_i(t)$ is a binary variable specifying whether project i is finished by time t. Since in most realistic problems the cumulative budget constraint is binding, that is there is never enough funding for all the available projects that are worth implementing, the optimized project sequence represented by the set of all t_is uniquely determines the schedule of projects (*1, 22*).

Optimization Method

A genetic algorithm (GA) is a search technique inspired by biologic natural selection and evolution: "survival of the fittest". Traditional techniques evaluate only one potential solution at a time when searching for the optimal solution, while a GA searches by concurrently examining a population of solutions. First, the GA generates many different solutions and computes their fitness value (which in most cases is the objective function value). Then, solutions are ranked based on their fitness value. Solutions with better fitness values are saved, while others are discarded. Some saved solutions are chosen as parents, and genetic operators, such as mutation and recombination operators, are applied on them to create a new generation of solutions. This process is repeated, until the specified number of generations is achieved or until the fitness function stops improving significantly. The GA includes the following steps (*23*):

1. Code the problem and determine the values of the parameters.
2. Form an initial population which contains n strings, where n depends on the type of problem examined. Evaluate the fitness function of every string.
3. Assuming the probability of choice is proportional to values of fitness function, choose n potential parents.
4. Randomly choose two or more parents and apply operators such as recombination and mutation operators to create offspring until a new population of n offspring is created.
5. Evaluate the fitness function for the new population for every offspring.
6. If the stopping criterion is reached, terminate the algorithm, and report the optimized solution (one with the best fitness value). Otherwise, return to step 3.

In this study, the initial population of the GA is generated randomly and solutions are represented by integer digits showing the sequence of the projects being implemented. Each individual in a population is defined as a string of numbers, each corresponding to a specific project in a sequence. The fitness function is the value of the objective function and is computed through the traffic assignment model.

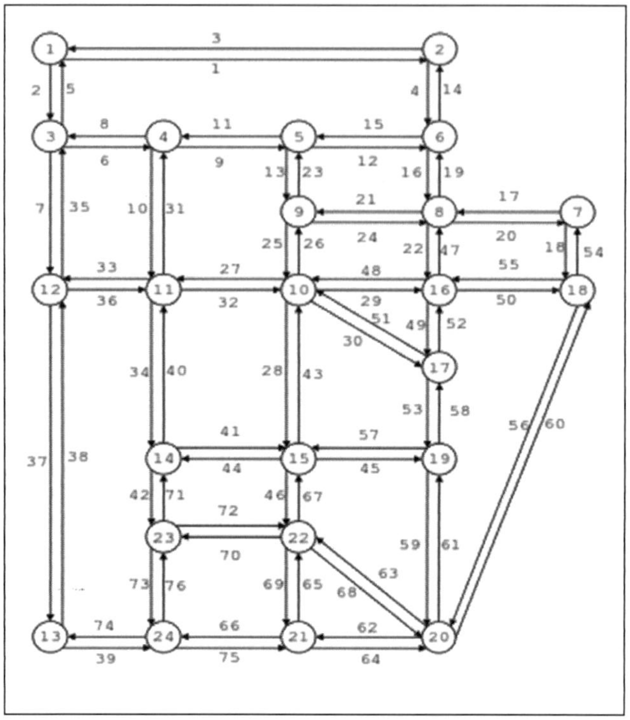

Figure 1. Graphical representation of the Sioux Falls network.

Case Study

The Sioux Falls network adopted from LeBlanc et al. (*24*) is used here as a case study. This network differs from the real network since it mainly includes the city's major arterials. It has been used in many previous studies. Figure 1 depicts the Sioux Falls network with 24 nodes and 76 links.

After running the traffic assignment model on the Sioux Falls network, links and intersections (nodes) that have critical volume-capacity (v-c) ratios are identified as an initial set of project improvements. The BPR function (*19*) used as a link performance function allows v-c ratios to exceed 1.0, which helps us identify the most congested links.

The project alternatives considered are link widenings (which are assumed to be applied symmetrically in both directions between the two connecting nodes because the O-D table is symmetric), and vertical, horizontal, or vertical and horizontal, improvements of intersections. Improvements are carried through the entire intersection

Table 1. Input Parameters

Parameter	Definition	Value and unit
c_i	Cost of lane addition	\$3 million/lane.mile
g	Demand growth rate	0.01/year
Ln_w	Lane widths	11 ft
B	Available budget	\$1.5 million/year
v	Value of time	\$15/veh.hr

for consistency with the number of lanes on the intersection's legs; there are two types of improvements that are considered in this paper: (i) N-S widening of the intersection between the North–South approaches, (ii) E-W widening of the intersection between East–West approaches. It is assumed that some projects should be bundled because it saves costs due to the joint use of resources and construction equipment. The assumption to bundle some projects is justified economically because it saves costs due to the joint use of resources and construction equipment.

In this example, it is assumed that the demand grows exponentially over the planning horizon as

$$d_{ij}^{t} = d_{ij}^{0*}(1+g)^{t} \tag{9}$$

where d_{ij}^{t} is the demand between origin i and destination j, d_{ij}^{0} is the base demand for the ij origin and destination (O-D) pair at time 0, and g is the growth rate per period. Some numerical values of the input parameters and their units are displayed in Table 1.

Nodes 8, 11 and 16 represent two-phased intersections in the Sioux Falls network. Intersections were modeled by adding one pseudo-link for each movement between link pairs, for example, for intersection 8, link 47 there are three pseudo-links (47002, 47004, 47006) for three movements (left-turn (002 part), through movement (004), and right-turn (006), respectively). Overall, for the three intersections (8, 11 and 16), 36 pseudo-links are added to the network. Table 2 shows the pseudo-links for intersections, their capacity, free flow travel time (t_0), and which pseudo-link belongs to which intersection. The capacity of each pseudo-link was set as the minimum value of the capacities of the two real links it connects.

Table 3 shows the initial volumes for each of the O-D pairs. It is evident that there are no trips originating and ending at nodes 8, 11, 16, because we consider them as intersections in the Sioux Falls network. In Table 4, the values of delay on intersection pseudo-links, the volumes on each pseudo-link, and pseudo v-c ratio are presented.

Table 2. Pseudo-Links for Intersections 8, 11, and 16, Their Capacity and Free Flow Travel Time

1st Node	2nd node	Pseudo-link ID	Capacity	t_0	Intersection
4	14	10004	731	2.652	11
4	10	10002	736	3	
4	12	10006	736	3.876	
6	16	16004	734	1.302	8
6	9	16006	734	1.302	
6	7	16002	734	1.302	
7	9	17004	757	1.5	8
7	6	17006	734	1.302	
7	16	17002	756	1.5	
8	17	22004	756	1.002	8
8	10	22006	728	2.7	
8	18	22002	756	1.614	
12	4	36002	736	3.876	11
12	10	36004	736	3	
12	14	36006	731	2.652	
14	12	40002	731	2.652	11
14	4	40004	731	2.652	
14	10	40006	731	2.652	
16	9	47002	756	2.892	8
16	6	47004	734	1.302	
16	7	47006	756	1.5	
17	10	52002	728	1.002	16
17	8	52004	756	1.002	
17	18	52006	756	1.002	
18	17	55002	784	1.002	16
18	10	55004	728	1.614	
18	8	55006	756	1.614	

Table 3. Baseline O-D Volumes for the Sioux Falls Network

O-D	1	2	3	4	5	6	7	8	9	10	11	12	13	14	15	16	17	18	19	20	21	22	23	24
1	0	200	120	360	180	240	300	0	360	840	0	180	360	180	300	0	300	120	180	180	60	240	180	120
2	200	0	6	18	6	30	12	0	18	36	0	12	18	6	12	0	18	6	6	12	6	12	6	6
3	120	6	0	18	6	18	6	0	12	18	0	18	12	6	6	0	6	0	6	6	6	6	6	6
4	360	18	18	0	30	30	30	0	48	72	0	42	36	30	30	0	30	6	18	24	12	24	30	18
5	180	6	6	30	0	18	12	0	48	60	0	12	12	12	18	0	18	6	12	12	6	12	12	6
6	240	30	18	30	18	0	24	0	24	48	0	18	18	12	18	0	36	6	18	24	6	18	12	6
7	300	12	6	30	12	24	0	0	36	114	0	48	30	18	30	0	60	60	30	36	18	36	12	6
8	0	0	0	0	0	0	0	0	0	0	0	0	0	0	0	0	0	0	0	0	0	0	0	0
9	360	18	12	48	48	24	36	0	0	168	0	42	36	36	60	0	60	12	30	42	24	42	36	12
10	840	36	18	72	60	48	114	0	168	0	0	126	114	132	240	0	234	42	108	156	78	162	108	54
11	0	0	0	0	0	0	0	0	0	0	0	0	0	0	0	0	0	0	0	0	0	0	0	0
12	180	12	18	42	12	18	48	0	42	126	0	0	84	42	48	0	42	12	18	30	24	48	42	30
13	360	18	12	36	12	18	30	0	36	114	0	84	0	36	42	0	36	6	24	42	36	78	48	48
14	180	6	6	30	12	12	18	0	36	132	0	42	36	0	45	0	42	6	24	30	24	72	66	24
15	300	12	6	30	18	18	30	0	60	240	0	48	42	45	0	0	90	18	48	66	48	156	60	30
16	0	0	0	0	0	0	0	0	0	0	0	0	0	0	0	0	0	0	0	0	0	0	0	0
17	300	18	6	30	18	36	60	0	60	234	0	42	36	42	90	0	0	42	102	102	42	102	36	18
18	120	6	0	6	6	6	60	0	12	42	0	12	6	6	18	0	42	0	24	100	6	24	6	6
19	180	6	6	18	12	18	30	0	30	108	0	18	24	24	48	0	102	24	0	78	30	78	24	12
20	180	12	6	24	12	24	36	0	42	156	0	30	42	30	66	0	102	100	78	0	78	150	42	30
21	60	6	6	12	6	6	18	0	24	78	0	24	36	24	48	0	42	6	30	78	0	114	42	36
22	240	12	6	24	12	18	36	0	42	162	0	48	78	72	156	0	102	24	78	150	114	0	132	72
23	180	6	6	30	12	12	12	0	36	108	0	42	48	66	60	0	36	6	24	42	42	132	0	4.8
24	120	6	6	18	6	6	6	0	12	54	0	30	48	24	30	0	18	6	12	30	36	72	4.8	0

Table 4. Values of Delay [sec/veh], Volume [veh] and v-c Ratios for Pseudo-Links at Intersections 8, 11, and 16

Intersection 8				Intersection 16				Intersection 11			
Pseudo-link	Delay	Volume	v-c	Pseudo-link	Delay	Volume	v-c	Pseudo-link	Delay	Volume	v-c
47002	17.97572	30.01127	0.039652	52002	27.89938905	479.8608	0.658934	40002	20.85631	222	0.303496
47004	21.64097	263.9938	0.359279	52004	19.50075055	142.3221	0.18804	40004	18.64281	83.99553	0.11483
47006	20.88573	228.6979	0.302161	52006	18.35787152	60.00031	0.079274	40006	19.53734	144	0.196862
24004	20.17247	186.009	0.245547	55002	20.00773625	178	0.2269	27002	21.30628	246	0.336306
24002	18.56341	78.01732	0.106177	55004	18.30268932	60.00649	0.0824	27004	21.94002	279.2566	0.379258
24006	17.97555	29.99829	0.039634	55006	19.52566035	144	0.190256	27006	17.542	0	0
17004	18.09034	108.0194	0.142594	22004	19.33305523	130.9015	0.17295	10004	18.52118	75.35613	0.103019
17006	19.00682	443.984	0.604234	22006	17.51455175	0	0	10002	17.542	0	0
17002	26.32533	203.9935	0.269521	22002	19.61527684	149.9971	0.19818	10006	19.91571	168.0003	0.228161
16004	20.46652	78.98595	0.107495	29002	18.08119079	43.67763	0.059977	36002	19.62855	150.0013	0.203716
16006	18.3156	60.00376	0.081661	29004	18.30256332	59.99731	0.082387	36004	22.87071	322.163	0.437529
16002	25.95317	433.1572	0.589499	29006	29.63127316	515.9972	0.708556	36006	20.85642	222.0059	0.303504

Figure 2 shows how the values of delay increase as the pseudo v-c ratio increases, for the three pseudo-links 36004, 16002, and 52002. These pseudo-links were chosen because of their large increases in delays as volume increases. The delay on each of the pseudo-links varies slightly, as can be seen in Figure 2.

Figure 3 shows overall intersection delay for the three intersections as function of the percentage of increase of the original O-D volumes, in 10% increments ranging from 10% to 140% of the original O-D table. In it the intersection delay usually increases as the percentage of volume increases, with intersections 8 and 16 having the greatest increases in delay. Due to traffic re-assignment, the delay increase is not monotonic at individual intersections.

Intersections 8 and 16 are considered for improvement based on their delay values. The links to be improved were chosen because of their high v-c ratios (above 0.6). Table 5 summarizes the list of projects. Intersections with the highest delay values and links with the highest v-c ratio are selected for improvement. Table 6 shows the bottleneck sequence and schedule of projects (ordered based on the projects' v-c ratios), greedy sequence and schedule (ordered based on their benefit-cost (b-c) ratio), and the GA-optimized sequence and schedule of projects. In this case, benefit is defined as the monetary value of total travel time savings from implementing one project, and cost is simply the implementation cost of each project. The present values of total costs after each project implementation are also shown in Table 6. These results indicate that the GA yields a better solution, that is, with lower total cost compared to sequences based on b-c and v-c ratios. This occurs because the GA process accounts for project interrelations, unlike common practices such as b-c ratio and congestion level rankings.

Figure 4 shows the performance of the GA; the optimized solution is reached after 22 generations. The stopping criterion for the GA was set at 10 successive similar solutions (shown in Figure 4) but, for more confidence in the results, we let it run further for 200 generations, which yielded the same solution. The CPU time for entire analysis is 3300 seconds. Table 7 demonstrates the sensitivity of the optimized sequence, schedule, and the objective function value (total cost) to changes in demand. Demand is changed by the same percentage for each cell in the O-D matrix. Table 7 also presents the sensitivity of results to changes in the available budget. The variation in budget is specified as different percentages of the original value, which was set to $1.5 million/year. It should be noted that unsteady budget flows do not increase the model's complexity or computation time.

Conclusion

The improvement of intersections and links in a network is just one example of interrelated alternatives for which

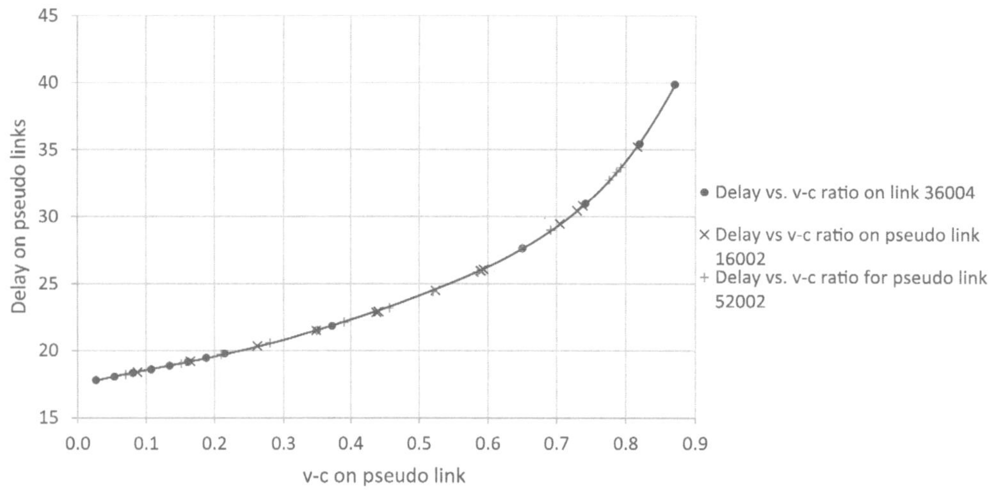

Figure 2. Delay vs. v-c ratio for pseudo-link 36004.

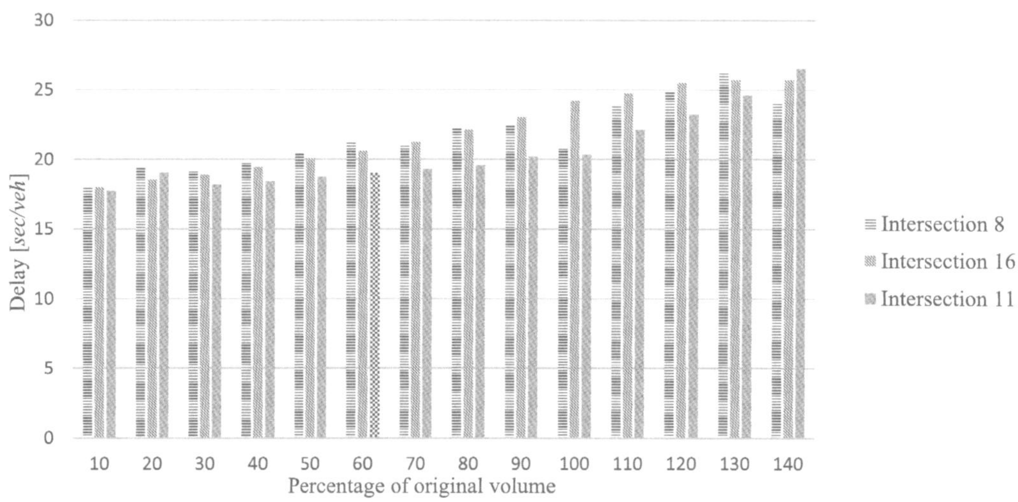

Figure 3. Intersection delay vs. percentage of the original volume of the O-D table.

the selections and scheduling of projects becomes a challenging optimization problem. This paper modifies the FW traffic assignment model to consider intersection flows and delays. This is done by introducing pseudo-links to the network and applying Akcelik's delay model. The modified model is then incorporated within a GA loop to optimize the selection and scheduling problem. Common prioritizing practices which are rankings based on b-c ratio and congestion level do not produce the optimal sequence of projects because they disregard the interrelations among projects, unlike the GA used here. This methodology can be applied more generally to other more complex cases. GAs can optimize very intractable objective functions without requiring restrictive assumptions about their structures which allows them to be efficiently combined with other evaluation tools, to solve selecting and scheduling problems.

Future research may focus on extending the model by incorporating more detailed evaluation methods (such as simulation models) to capture dynamic effects in

Table 5. List of Candidate Projects and Their Costs

Project ID	Project description	Project cost
1	Improvement of links 69 & 65	$1,800,000.00
2	Improvement of links 30 & 51	$4,800,000.00
3	Improvement of links 62 & 64	$3,900,000.00
4	Improvement of links 68 & 63	$4,200,000.00
5	horizontal improvement of intersection 8 (pseudo-links 17, 24)	$220,880.00
6	vertical improvement of intersection 8 (pseudo-links 16, 47)	$220,880.00
7	horizontal improvement of intersection 16 (pseudo-links 29, 55)	$220,880.00
8	vertical improvement of intersection 16 (pseudo-links 22, 52)	$220,880.00

Table 6. Bottleneck, Greedy, and GA-Optimized Schedule of Projects with Corresponding Total Costs

Bottle neck sequence	Present value of total cost ($)	Schedule (years)	Greedy order sequence	Present value of total cost ($)	Schedule (years)	GA- optimized sequence	Schedule (years)	Present value of total cost ($)
1	532,870,256	1.2	2	1,295,746,013	3.2	7	0.15	68,455,247
2	1,701,652,369	4.4	3	2,095,244,334	5.8	6	0.29	135,967,546
3	2,421,799,340	7	6	2,134,903,906	5.95	2	3.49	1,396,017,082
4	3,025,641,095	9.8	7	2,174,285,216	6.09	3	6.09	2,174,056,728
5	3,053,349,907	9.95	8	2,213,386,599	6.24	4	8.89	2,825,298,019
6	3,080,981,763	10.09	5	2,252,222,345	6.39	1	10.09	3,057,265,698
7	3,108,535,464	10.24	4	2,902,828,902	9.19	5	10.24	3,085,002,834
8	**3,136,038,746**	10.39	1	**3,134,797,942**	10.39	8	10.39	**3,112,075,181**

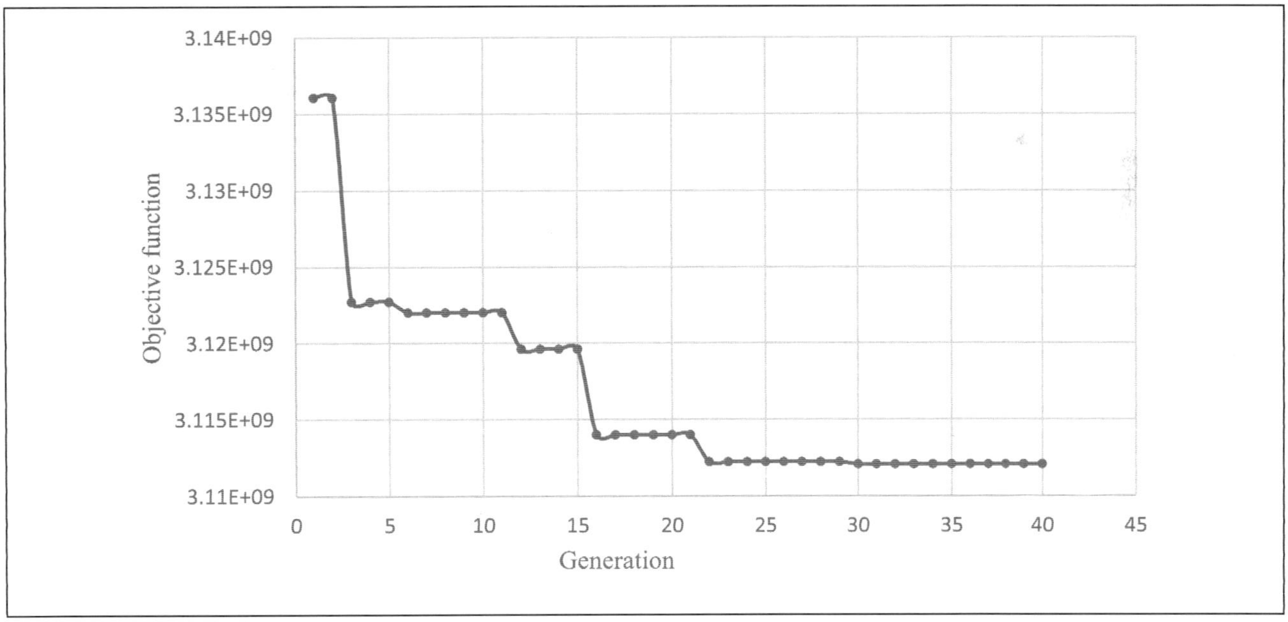

Figure 4. Performance of the genetic algorithm.

congested networks that are missed by the FW algorithm. Future model versions may also consider more elaborate intersection configurations, control policies and cyclical variations in daily and weekly traffic. GAs could be solved considerably faster by distributing the evaluation of population members among multiple

Table 7. Sequence and Schedule Change Subject to Demand Change and Sequence and Schedule Change Subject to Budget Change

	Sequence	Schedule (years)	Present value of total cost
Demand			
50%	6-7-3-2-4-1-8-5	0.1-0.3-2.9-6.1-8.9-10.1-10.2-10.4	$1,532,306,913
75%	7-6-2-4-3-1-5-8	0.1-0.3-3.5-6.3-8.9-10.1-10.2-10.4	$2,309,379,871
100%	7-6-2-3-4-1-5-8	0.1-0.3-3.5-6.1-8.9-10.1-10.2-10.4	$3,112,075,181
110%	7-6-2-4-3-1-8-5	0.1-0.3-3.5-6.3-8.9-10.1-10.2-10.4	$3,452,620,779
125%	7-6-2-3-4-1-8-5	0.1-0.3-3.5-6.1-8.9-10.1-10.2-10.4	$3,989,723,752
Budget			
50%	6-7-2-4-3-1-8-5	0.3-0.6-7-12.6-17.8-20.2-20.5-20.8	$4,272,866,372
66%	6-7-2-4-3-1-5-8	0.2-0.4-5.2-9.4-13.3-15.1-15.4-15.6	$3,832,339,985
100%	7-6-2-3-4-1-5-8	0.1-0.3-3.5-6.1-8.9-10.1-10.2-10.4	$3,112,075,181
133%	7-6-3-2-4-1-5-8	0.1-0.2-2.2-4.6-6.7-7.6-7.7-7.8	$2,594,955,256
200%	7-6-3-2-4-1-5-8	0.07-0.1-1.4-3-4.4-5-5.1-5.2	$1,935,240,006

processors. Moreover, individual improvements (resurfacing, widening) could be grouped to form a project, bus traffic could be traced along with passenger vehicles in the traffic assignment method, and different cost rates could be assumed for different types of improvements implemented.

Acknowledgments

The authors are grateful for the support received from the National University Transportation Center at the University of Maryland, which partly funded this work, and from TRB reviewers who provided valuable suggestions and comments.

References

1. Shayanfar, E., A. S. Abianeh, P. Schonfeld, and L. Zhang. Prioritizing Interrelated Road Projects Using Metaheuristics. *Journal of Infrastructure Systems*, Vol. 22, No. 2, 2016, p. 04016004.
2. Wardrop, J. G. Road Paper. Some Theoretical Aspects of Road Traffic Research. *Proceedings of the Institution of Civil Engineers*, Vol. 1, No. 3, 1952, pp. 325–362.
3. Webster, F. V. *Traffic Signal Settings*. Road Research Technical Paper No. 39. Road Research Laboratory, London, 1958.
4. *Highway Capacity Manual*. TRB, National Research Council, Washington, D.C., 2000.
5. Akcelik, R. The Highway Capacity Manual Delay Formula for Signalized Intersections. *ITE Journal*, Vol. 58, No. 3, 1988, pp. 23–27.
6. Akcelik, R. *Appendix: A Note on the Generalized Delay Model*. Compendium of Technical Papers, 60th Annual Meeting of Institute of Transportation Engineers, Orlando, Fla., 1990, pp. 29–32.
7. Teply, S. Quality of Service in the New Canadian Signal Capacity Guide. *Proc., International Symposium on Highway Capacity and Level of Service*, A. A. Balkema, Rotterdam, 1991, pp. 377–386.
8. Vincent, R.A., Mitchell, A.I., Roberston, D.I. *User Guide to TRANSYT Version 8*. Transportation and Road Research Laboratory (TRRL), Workingham, Berkshire United Kingom, 1980.
9. Heidemann, D. Queue Length and Delay Distributions at Traffic Signals. *Transportation Research Part B: Methodological*, Vol. 28, No. 5, 1994, pp. 377–389. https://doi.org/10.1016/0191-2615(94) 90036-1
10. Olszewski, P. S. Modeling Probability Distribution of Delay at Signalized Intersections. *Journal of Advanced Transportation*, Vol. 28, No. 3, 1994, pp. 253–274. https://onlinelibrary.wiley.com/doi/abs/10.1002/atr.5670280306
11. Weingartner, H. M. Capital Budgeting of Interrelated Projects: Survey and Synthesis. *Management Science*, Vol. 12, No. 7, 1966, pp. 485–516. https://doi.org/10.1287/mnsc.12.7.485
12. Cochran, M. A., E. B. Pyle, L. C. Greene, H. A. Clymer, and A. D. Bender. Investment Model for R&D Project Evaluation and Selection. *IEEE Transactions on Engineering Management*, Vol. EM-18, No. 3, 1971, pp. 89–100. https://dx.doi.org/10.1109/TEM.1971.6447136
13. Nemhauser, G. L., and Z. Ullmann. Discrete Dynamic Programming and Capital Allocation. *Management Science*, Vol. 15, No. 9, 1969, pp. 494–505. https://doi.org/10.1287/mnsc.15.9.494
14. Mehrez, A., and Z. Sinuany-Stern. Resource Allocation to Interrelated Projects. *Water Resources Research*, Vol. 19, No. 4, 1983, pp. 876–880. http://dx.doi.org/10.1029/WR019i004p00876
15. Florian, M., and S. Nguyen. An Application and Validation of Equilibrium Trip Assignment Methods. *Transportation Science*, Vol. 10, No. 4, 1976, pp. 374–390. https://doi.org/10.1287/trsc.10.4.374
16. Sheffi, Y. *Urban Transportation Networks: Equilibrium Analysis with Mathematical Programming Methods*. Prentice-Hall, Inc., Englewood Cliffs, N.J., 1985.
17. Ran, B., and D. Boyce. *Dynamic Urban Transportation Network Models. Lecture Notes in Economics and Mathematical Systems*. Springer-Verlag, Berlin, Heidelberg, Vol. 417, 1994.

18. Ran, B., and D. Boyce. *Modelling Dynamic Transportation Networks: An Intelligent Transportation System Oriented Approach*. Springer-Verlag, Berlin, Heidelberg, 1996.
19. Frank, M., and P. Wolfe. An Algorithm for Quadratic Programming. *Naval Research Logistics (NRL)*, Vol. 3, No. 1–2, 1956, pp. 95–110. http://dx.doi.org/10.1002/nav.3800030109
20. Van Leeuwen, J. *Handbook of Theoretical Computer Science (Vol. A): Algorithms and Complexity*. The MIT Press/Elsevier, Cambridge MA/Amsterdam, The Netherlands, 1991.
21. Dorigo, M., and T. Sttzle *Ant Colony Optimization*. The MIT Press, Cambridge, Mass., 2004.
22. Jong, J. C., and P. Schonfeld. Genetic Algorithm for Selecting and Scheduling Interdependent Projects. *Journal of Waterway, Port, Coastal, and Ocean Engineering*, Vol. 127, No. 1, 2001, pp. 45–52.
23. Teodorovic, D. *Transportation Networks*. University of Belgrade, Faculty of Transport and Traffic Engineering, Belgrade, 2007.
24. LeBlanc, L. J., E. K. Morlok, and W. P. Pierskalla. An Efficient Approach to Solving the Road Network Equilibrium Traffic Assignment Problem. *Transportation Research*, Vol. 9, No. 5, 1975, pp. 309–318. https://doi.org/10.1016/0041-1647(75)90030-1

The Standing Committee on Transportation Programming and Investment Decision-Making (ADA50) peer-reviewed this paper (18-05088).

TRR
JOURNAL OF THE TRANSPORTATION RESEARCH BOARD

Article

Transportation Research Record
2018, Vol. 2672(51) 12–19

DOI: 10.1177/0361198118756632
journals.sagepub.com/home/trr
SAGE

Transportation Planning for Connected Autonomous Vehicles: How It All Fits Together

Bobby J. Cottam[1]

Abstract

As connected and autonomous vehicle (CAV) technology continues to evolve and rapidly develop new capabilities, it is becoming increasingly important for transportation planners to consider the effects that these vehicles will have on the transportation network. It is evident that this trend has already started; over 60% of long-range transportation plans in the largest urban areas now include some discussion of CAVs, up from just 6% in 2015. There are also numerous CAV pilot programs currently underway that entail testing vehicle-to-vehicle (V2V) and vehicle-to-infrastructure (V2I) interaction in both isolated and real-world environments. In this review of the current assessments for CAV impacts, two primary trends are identified. First, there is a great deal of uncertainty that is not being explicitly considered and properly accounted for in the transportation-network planning process. Second, the predictions that are being made are not considering potential policy or planning actions that could shape or affect the impacts of CAVs. This paper provides a picture of how ongoing CAV research interacts with current transportation planning practices by examining how the methods, the ranges of predictions, and the different sources of uncertainty in each method impact the planning process and potential system outcomes. Finally, it will identify best practices from decision analysis to help plan the best possible future transportation networks.

In recent years, connected and autonomous vehicles (CAVs) have become an increasingly discussed topic. Their potential impact is currently a point of much debate, with opinions ranging from them being the savior of the transportation industry to the beginning of traffic doomsday scenarios. With such a potentially revolutionary technology rapidly approaching, it is important for transportation agencies to begin planning for CAVs. With so much uncertainty and so many different agencies involved, it can be difficult to create meaningful plans. To create a coherent strategy that will provide the most beneficial impact, it is important to understand the many different tools available for planning. This paper reviews the literature associated with the current state of transportation planning for CAVs. Rather than provide a comprehensive review of any single topic related to CAVs, it seeks to show the range of research on each topic sufficient for the reader to understand the sources of uncertainty and the relationships between each topic. The paper begins by unpacking the acronyms and terminology associated with this field. Then the current language, if any, that is used to discuss and plan for CAVs is examined. This examination considers metropolitan planning organizations (MPOs) with a population over one million and the current status of the existing plans in their organizations. These organizations have been charged with maintaining the long-range plans and the distribution of federal money for these highly populated areas. Next the current techniques in academia and industry for estimating the impacts of CAVs is examined and the range of research results for their impact on major travel characteristics such as travel time, travel demand, capacity, and land use is provided. The current test beds and existing pilot programs that are being conducted, what data can be derived from them, and how that data will allow planners and researchers to gain a better understanding of the impacts of CAVs are summarized. The final section gathers the conclusions that can be drawn based on this review, outlines the major sources of uncertainty, and discusses what work still needs to be done ensure an efficient integration of CAV technology into the transportation systems of the future.

[1]Department of Industrial Engineering, Bell Engineering Center, University of Arkansas, Fayetteville, AR

Corresponding Author:
Address correspondence to Bobby J. Cottam: bjcottam@uark.edu

Level	Name	Narrative definition	Execution of steering and acceleration/ deceleration	Monitoring of driving environment	Fallback performance of *dynamic driving task*	System capability (*driving modes*)	BASt level	NHTSA level
***Human driver* monitors the driving environment**								
0	No Automation	the full-time performance by the *human driver* of all aspects of the *dynamic driving task*, even when enhanced by warning or intervention systems	Human driver	Human driver	Human driver	n/a	Driver only	0
1	Driver Assistance	the *driving mode*-specific execution by a driver assistance system of either steering or acceleration/deceleration using information about the driving environment and with the expectation that the *human driver* perform all remaining aspects of the *dynamic driving task*	Human driver and system	Human driver	Human driver	Some driving modes	Assisted	1
2	Partial Automation	the *driving mode*-specific execution by one or more driver assistance systems of both steering and acceleration/deceleration using information about the driving environment and with the expectation that the *human driver* perform all remaining aspects of the *dynamic driving task*	**System**	Human driver	Human driver	Some driving modes	Partially automated	2
***Automated driving system* ("system") monitors the driving environment**								
3	Conditional Automation	the *driving mode*-specific performance by an *automated driving system* of all aspects of the *dynamic driving task* with the expectation that the *human driver* will respond appropriately to a *request to intervene*	System	**System**	Human driver	Some driving modes	Highly automated	3
4	High Automation	the *driving mode*-specific performance by an *automated driving system* of all aspects of the *dynamic driving task*, even if a *human driver* does not respond appropriately to a *request to intervene*	System	System	**System**	Some driving modes	Fully automated	3/4
5	Full Automation	the full-time performance by an *automated driving system* of all aspects of the *dynamic driving task* under all roadway and environmental conditions that can be managed by a *human driver*	System	System	System	**All driving modes**	-	

Figure 1. SAE levels of autonomy (2).

CAV Terminology

To begin it is necessary to unpack the terminology associated with this technology. Intelligent transportation system (ITS) is a very broad term used to discuss advanced technology and computing power being applied to transportation systems. ITS includes such basic instruments as variable message signs on highways and also encompasses tools as complicated as CAVs. Autonomous vehicles (AVs) refer to vehicles that can drive themselves with no input from a human driver. Such a vehicle possesses full longitudinal and lateral control. The Society of Automotive Engineers (SAE) has established a scale from 0 to 5 (Figure 1) where 0, 1, and 2 are various degrees of driver-assistance programs such as adaptive cruise control and lane-keeping assistance programs. The United States Department of Transportation (USDOT) as well as the NHTSA *Federal Automated Vehicle Policy* uses the term highly automated vehicle (HAV) for SAE levels 3 to 5 as those automated vehicle systems "are responsible for monitoring the driving environment" (*1*). The distinction between levels three through five relate to the safety fallback state and the operating environment of the vehicle.

Connected vehicles refer to vehicles that can communicate with other objects on the roadway. Broadly, this is termed V2X. However, this term can be modified to be more specific. When communicating with another vehicle, the technology can be referred to as V2V; when it is communicating with infrastructure objects, it can be referred to as V2I. There are many different communication media that are being considered for this task. One of the most commonly discussed methods is dedicated short-range communication (DSRC). In 2003 the Federal Communications Commission (FCC) allocated bandwidth in the 5.9 GHz band specifically for ITS use (*3*). Nokia Advanced 5G wireless networks are also claiming they will have the required reliability and low latency required for CAV operation (*4*). The emerging internet of things (IoT) is often discussed alongside connected vehicle technologies. The IoT refers to the interconnectedness of many different "smart" devices. "Smart cities" is a term generically applied to the new ideal of a city that employs a host of technological solutions, including ITS technologies, to greatly improve the city as a whole.

State of MPO Planning

One way to understand trends in transportation planning across the country is to examine the reports being produced by some of the largest MPOs. MPOs are governmental organizations that are responsible for urbanized areas containing over 50,000 people. Currently there are over 400 MPOs across the country, and these MPOs account for nearly 75% of the U.S. population (*5*). They are required to maintain a transportation plan that looks at least 20 years into the future. These are typically referred to as either a regional transportation plan (RTP) or a long-range transportation plan (LRTP). These plans set the vision and the tone for the entire region's development and transportation investment. Seeing the extent

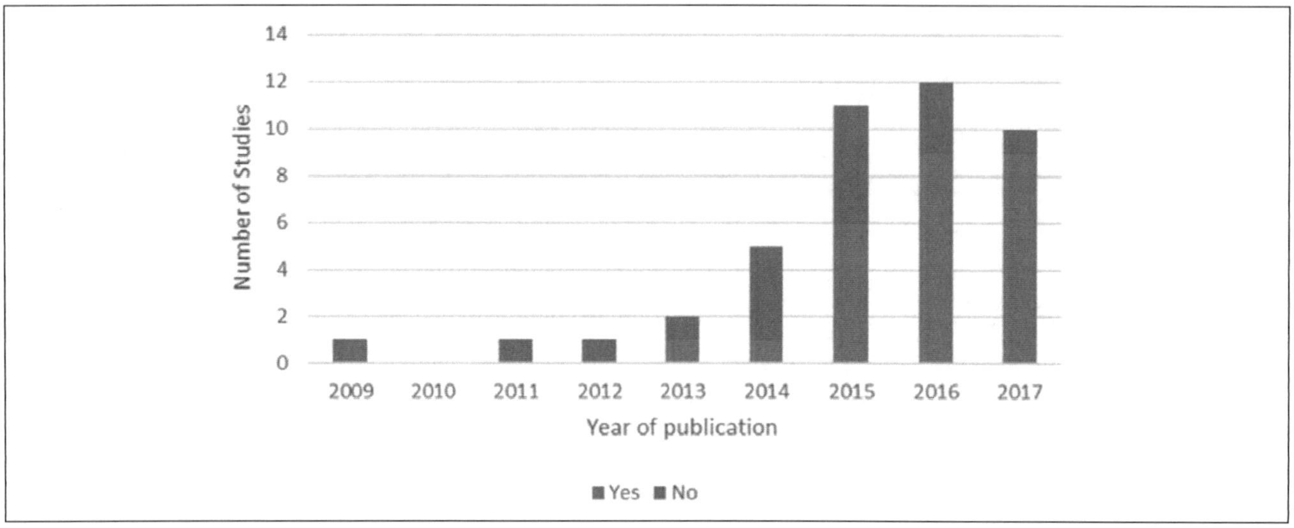

Figure 2. Number planning documents discussing CAVs.

to which these plans, if any, discuss CAVs is very important in understanding how organizations are beginning to plan for their integration. This paper reviewed all MPOs with a 2010 urbanized area population over 1 million (44 in total) to determine how, if at all, they are planning for AVs (*5*). According to a study by the National League of Cities done in 2015, only 6% of plans (from 68 of the largest cities in the U.S.) were considering the potential effects of driverless technology (*6*). However, at the time of this review, only 2 years later, 61% of MPOs had discussed CAVs in one of their primary planning documents. This number would be even higher if we were to consider the state-level plans that cover some of the remaining 16 MPOs. This significant increase demonstrates how rapidly CAVs are gaining traction. Only two out of the ten planning documents that were accepted prior to 2015 considered CAVs. Since LRTPs are updated every four to 5 years, this means that approximately 50% of the plans have been updated in the 2.5 years since the league of cities study, and nearly all of those have been altered to include a discussion of CAVs. This incredibly high rate of inclusion over such a short time indicates just how quickly this technology has captured the attention of planners. Figure 2 shows the number of planning documents that do and do not discuss CAVs based on publication year. Only 43 documents are shown in as two MPOs had a joint regional planning document that discussed CAVs.

Not all of the plans gave the CAVs equal discussion. Some plans simply state that CAVs are an emerging technology that needs to be monitored, such as in the Orange County Transportation Authority's LRTP "Outlook 2035" (*7*). Other plans contained a much more thorough review of the technology and discussed approaches on how to best capitalize on it. The most advanced plans incorporated CAVs into scenario planning and considered the drivers of technology and the potential impacts. The Baltimore Regional Transportation Board went so far as to conduct focus groups to assist with the scenario planning (*8*). Another of the most extensive reviews was performed by the Miami-Dade Transportation Planning Organization. They developed an *Impact of Future Technology in the 2045 LRTP Report* (*9*). One of the key points was that "it cannot be stressed enough that a new suite of travel demand models (TDM) must be in place to develop the 2045 LRTP to address the impact of autonomous/connected vehicles." The Orlando ITS master plan also made numerous references to CAVs. It also discusses the needs associated with the Tampa Hillsborough County Expressway Authority (THEA) pilot project, which intends to equip "approximately 10 buses, 10 streetcars and 1,500 participants' automobiles with wireless communication devices that can exchange traffic and safety information with other vehicles and with roadway infrastructure in Downtown Tampa" (*10*).

Impact Estimates of CAVs on Transportation Systems

We can see that many agencies are discussing CAVs and beginning to plan for them, but at the LRTP level the impacts were broadly discussed. More emphasis is needed on understanding what aspects of the transportation network are likely to be affected, determine if those impacts will be positive or negative, and what policy level decisions the MPO could enact to encourage or discourage the effects. This section will look more closely at the predicted CAV impacts to understand the range of predictions and the methods and data that were used. At the most basic level of transportation analysis, two elements determine whether or not a road will experience congestion: capacity and demand. Together these two elements allow for the analysis of the volume to capacity ratio that is a key metric for understanding roadway performance.

Impact of CAVs on Capacity Estimates

Capacity estimates can be derived in two ways, a high-level, conceptual estimate or a detailed analysis using modeling and simulation. The first is a more philosophical method. One considers the differences between a human driver and a computer driver. The impact of those differences is then considered, and an estimate is given based on the elimination of the human element and the inclusion of computer element. In general, these estimates provide upper bounds on the probable benefits from CAVs as they are based on the assumptions of 100% market penetration and nearly flawless computer systems.

Safety and accident-reduction estimates are often handled conceptually. The commonly cited figure, that AVs could reduce accidents by over 90%, cited the NHTSA report that attributed human causes to 93% of traffic accidents (*11*). These human causes included, among others, distracted driving (e.g., cell phone and rubbernecking) and alcohol related incidents. Rubbernecking, which is responsible for a reduction in traffic the opposite direction of an accident, would also be eliminated by AVs. The reduction in capacity by rubberneckers, who slow down to see what happened, can be 5% to 25% of the drop-in capacity caused by the accident in the primary direction of travel (*12*). With traffic incidents causing approximately 25% of traffic delays on the roads this would be a substantial improvement (*13*).

Looking at traffic from a per lane basis rather than as an integrated whole, one finds predictions such as a highway-capacity increase of 100%. Farmer proposes CAVs would make this possible by reducing perception and reaction time, allowing a corresponding reduction in the safe following distance from 1.5 s down to 0.1 s (*14*).

The modeling and simulation method allows for more in-depth analysis and interaction of the variables. This allows the researcher to more precisely understand the effects of CAVs in specific scenarios and provides a better understanding of the effects of different market-penetration rates. In a highway setting, connective adaptive cruise control (CACC) and its effect on car-following models has been a primary research focus for determining the changes in lane capacity from CAVs. Using simulation of a one-lane highway segment, Shladover et al. found that the lane capacity increased quadratically with market penetration (*15*). There was negligible growth in capacity at 10% market penetration, a 20% increase in lane capacity at 50% market penetration, and the capacity nearly doubled at 100% market penetration. This trend became more linear when it was assumed that vehicles not equipped with CACC were still able to broadcast their position and velocity to assist the CACC equipped vehicles, a change the authors suggested could be a very inexpensive aftermarket upgrade. That could then translate into a program that could be supported by a planning agency. Talebpour and Mahmassani found similar results of approximately 100% increase in throughput on highways for 100% CAVs when mathematically analyzing a single-lane highway with a merge segment using an analysis of string stability (*16*). It is worth noting, however, that neither of these methods appears to take into account the weaving behavior, specifically the interaction between CAVs and human-driven vehicles. Within either homogenous group, there are advantages to the interactions. Two automated drivers will be able to coordinate a precise lane change reaction even during heavy traffic. Two human drivers can gain advantages in heavy traffic by weaving a vehicle in and out or flashing their headlights to let another driver know they will yield even though the rules of the road would not assume that behavior. When there is an automated and a human driver engaging in these lane changes, neither benefit will be present, which could reduce the capacity of the roadway in a manner that is not directly obvious by only considering the lane capacity.

A report from the UK DOT used microsimulation to look not only at different levels of vehicle automation and market penetration, but also at the different performance characteristics of CAVs based on occupant's comfort. The results showed that the assertiveness of the CAVs would dramatically impact their effect on roadway capacity even to the extent of decreased performance measures for some scenarios (*17*). These may be some of the most important findings from all of the simulations that have been done. It illustrates the importance of the model assumptions. With very little input so far from original equipment manufacturers (OEMs) on design specifications, no existing regulations, and no large-scale, real-world data, planners face difficult challenges. For example, the Texas DOT best-practices guidebook for preparing Texas for CAVs states that since "traffic models of AVs are recent developments in the literature, our car-following model may not be a completely accurate model of AV behavior when AVs become available to most travelers" (*18*).

Later in the best-practices guidebook, extensive modeling of safety behavior showed a dramatic reduction in rear-end collisions, but minimal reduction of conflicts at intersections or lane changes. With existing technology for blind spot detection, vehicles' ability to communicate their locations, and research on the ability of CAVs to eliminate traffic signals, it would seem that both the conflicts and accidents at intersections or lane changes should decrease, especially at full market penetration. No noticeable reduction indicates that either the driving behavior in the models is still in need of some adjustment, or that what is determined as a possible conflict by the FHWA surrogate safety-assessment model needs to be adjusted to account for the more precise driving enabled with CAVs.

Impacts of CAVs on Demand Estimates

CAVs will have the ability to significantly alter how people travel. These changes, even more so than the capacity changes, are hard to estimate. For example, if on a particular

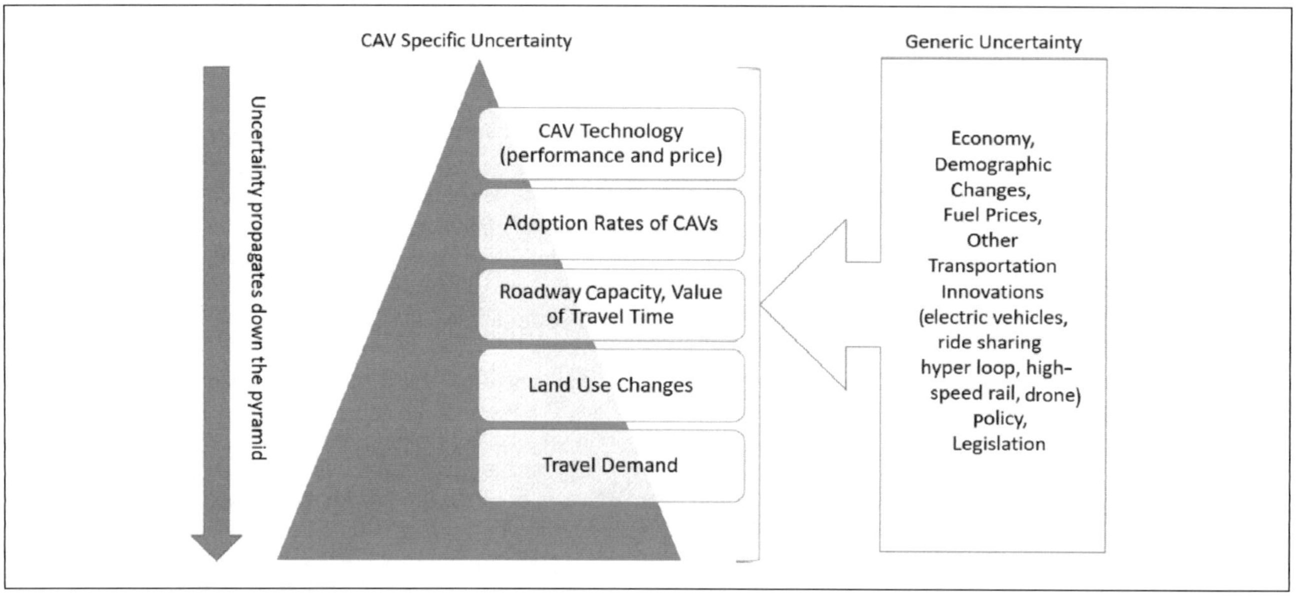

Figure 3. Uncertainty in transportation planning for CAVs.

highway it is expected that CAV technology will create a reduction in travel time of five to fifteen minutes, and a five-minute reduction in travel time will increase travel on that roadway by 20% to 60% while a fifteen-minute reduction in travel time would be likely to induce 50% to 120% more traffic, then the true variability in demand is now actually 20% to 120%, much larger than either individual estimate would have predicted in isolation. Figure 3 depicts the uncertainties specific to CAVs, how they propagate through to different parts of the planning process, and the general uncertainty that has always affected transportation planning. Each step down the pyramid must consider both the uncertainty of every item above it and the general uncertainties when estimating the possible impacts.

Putting this uncertainty aside for a moment, let us consider the research being done on the impacts of CAVs on travel demand. The Miami-Dade Transportation Planning Agency stated that it is crucial to update their travel-demand models before the next update of their LRTP (*9*).

Fagnant and Kockelman proposed that vehicle miles traveled (VMT) could rise up to 26% when AV market penetration reaches 90%, based on the latent demand in the network (*19*). It is also a common assumption that VMT will increase due to a decreased value of time spent traveling, because time in the car can now be used to accomplish other tasks besides driving. However, with both of these changes, it is important to understand their potential impact. If delay and congestion in a downtown business district are the primary concern, is it truly important that VMT has increased as a result of sprawl and commutes from commuters' farther out in the suburbs? There may be increased VMT, but that is not going to contribute to increased delay or congestion in the downtown area. The same number of trips are coming into the downtown area; they are simply coming from farther away. If, however, the goal is to minimize emissions, then this is still very problematic as increased mileage, all other things being equal, will increase emissions. Thus, it is important for planners to clearly specify their objectives and goals before trying to evaluate any possible courses of action.

Some individual road improvement projects are also beginning to look at the impact of CAVs. Given that predicting both the uncertainty and the demand has been so difficult, some projects have tried to look at the problem from the other direction. Instead of predicting the uncertainty, they perform a sensitivity analysis on several of the best alternatives to see for what ranges of demand the design solution remains the optimal choice. This change in thinking means a planner or decision maker no longer must determine a specific value for the change in demand but simply how likely it is that the demand will shift beyond the determined thresholds.

The entire transportation system is interconnected, and CAVs will have impacts beyond passenger vehicle travel. One of these impacts is the effect CAVs will have on transit. Will CAVs act as a replacement for transit and thereby reduce transit riders, or will it act as an enabler to transit by providing first- and last-mile service, thereby expanding the population served by every transit stop location? Policy and investment can help to shape that answer. The entire concept and structure of parking is also in question with CAVs—another area that can be heavily influenced by transportation planning and urban design. What will happen to land that was once needed for parking? Will there be more streamlined processes for pick-up, drop-off, and loading zones? This technology is not simply limited to automobiles either;

Virginia has signed a signed a bill that will allow autonomous delivery robots on sidewalks. The robots will be allowed to travel at up to 4 mph and can be used for small deliveries such as lunch, groceries, and medicine (*20*). With automated trucks and drone delivery being considered the standard, predictive models for shipping and commerce-related travel might start to look very different as well.

Because of the difficulty in accurately predicting the most likely scenario, an FHWA report advocated for the development of alternate futures for scenario analysis, to look at the different possible ways that the technology could develop (*21*). Looking at alternative futures, scenario analysis, using performance measures to assess the success of different objectives as discussed in *Performance Based Planning and Programming Guidebook* (*22*), and dealing with significant uncertainty is the perfect case to use decision-analysis techniques. Multiple objective decision analysis (MODA), which is considered a best practice in the field of decision analysis, is able to handle all of these aspects of a complicated decision. A fast summary of the methods for decision analysis can be found in "Decision Analysis in One Chart" (*23*) with a more detailed description of the techniques found in *The Decision Analysis Handbook* (*24*). MODA allows decision makers to compare the value provided across conflicting objectives such as safety, emissions, and delay, to find the ideal alternative to provide the maximum value to the stakeholders in a way that is transparent and traceable.

CAV Pilot Programs

CAV pilot programs are a crucial step in reducing much of the uncertainty in determining capacity and demand. The programs provide a crucial step into the calibration and validation of many of the models that have been developed. Many cities have reached independent agreements with various technology companies to enable AVs to operate on their streets. With the support, in the form of a 2015 executive order, Arizona became a favorable location for testing of AVs (*25*). Uber took advantage of this and began testing their AV fleet on Arizona roads at the beginning of this year. Boston, Massachusetts has entered into an agreement with NuTonomy allowing them to test their AVs on a small portion of their roadway system (*26*). These programs will help advance the state of AV development. It is unclear how much data will be available from a research standpoint, as all of these companies have a vested interest in protecting their intellectual property. Additionally, the small number of vehicles in these programs means there will be little if any useful data produced about the macro effects on network performance characteristics, a concern of many regional planning models. The best case would likely be to receive some operational performance data, such as following distance, reaction times, and so forth, and to gauge the public response to these vehicles to better understand public perception and how that could influence the adoption rates.

A more promising set of pilot programs from a data standpoint may be from the ten proving grounds that the DOT announced in January (*27*):

1. City of Pittsburgh and the Thomas D. Larson Pennsylvania Transportation Institute
2. Texas AV Proving Grounds Partnership
3. U.S. Army Aberdeen Test Center
4. American Center for Mobility (ACM) at Willow Run
5. Contra Costa Transportation Authority (CCTA) & GoMentum Station
6. San Diego Association of Governments
7. Iowa City Area Development Group
8. University of Wisconsin-Madison
9. Central Florida Automated Vehicle Partners
10. North Carolina Turnpike Authority

The information that has been released on these programs is very promising. The Texas AV Proving Grounds Partnership explained one of their locations for real-world highway testing (*28*).

"Highway: IH-30 from IH-35W in Fort Worth to IH-35E in Dallas is a modern multi-lane expressway. It includes a 10-mile, reversible, protected, managed lane facility operated by TxDOT, which can be closed during off-peak periods. Thus, AV testing on IH-30 can start in the protected environment of the closed managed lanes, followed by testing in the relatively stable conditions of the managed lanes with regular traffic, and finally, testing in the general purpose lanes."

This type of testing environment could provide enormous amounts of data for calibrating microsimulation models. Many of the string stability and capacity tests could be verified, and the varying traffic volumes would be helpful in understanding how the behavior and interactions with CAVs differ between free-flow conditions and congested conditions.

The Wisconsin Automated Vehicle Proving Grounds has among its stated objectives: testing and validation of methods for AV systems, human–machine interfaces and responses, and passenger comfort (*29*). These perception-based aspects are as important as mechanical and technical capabilities. As consumers become more comfortable with the technology they will probably become more comfortable with characteristics that could increase transportation-systems performance, such as higher speeds and closer following distances.

While many of these programs so far have placed a stronger emphasis on the autonomous rather than connected aspect of CAVs, there are also several programs that are beginning massive deployments of V2I systems. New York City will equip 10,000 vehicles with V2V communication devices and 250 signals with roadside equipment to communicate with those vehicles and provide them with additional information (*30*). The Columbus smart-city program will be installing V2V devices in travelers' cars and will also begin employing electric CAVs to provide first- and last-mile trips around a commercial retail hub (*31*). Because the V2V device installment program has voluntary enrollment, it will be able to provide additional adoption rate information. Depending on the data collection associated with the V2V devices, it might also be possible to gain before and after data on travel-time reduction and a picture of how that data changes through time with additional users. Finally, the first- and last-mile ride services may provide valuable insight about the potential of leveraging CAVs to increase transit ridership, rather than the decrease in transit ridership that has been predicted with the rise of CAVs.

Conclusion

Planning for CAVs has made dramatic progress in the last several years. It is being researched, discussed, and planned for, and the body of knowledge is growing at an exponential rate. Initial methods are currently being developed to predict the impact of these vehicles, but the uncertainty that remains is still staggering. It is as if planners are being asked to predict and plan for airports after the Wright brothers inaugural flight. The ability to change how people and goods are transported is being revolutionized, but the development is happening at a more rapid pace than ever before in history. With every facet of this technology, there are associated uncertainties and all of those uncertainties interact in an additive, if not multiplicative, fashion. Understanding the implication of how any individual piece of research fits into the greater planning context is crucial. Since planning for CAVs cannot wait until this uncertainty is resolved, plans must instead address the issues caused by uncertainty in the most direct and useful fashion possible. This review suggests four important action steps that planners can take to manage this uncertainty.

First, be transparent with assumptions related to CAV performance and penetration—directly stating any assumptions or simplifications that were made, and what impact those might have, so when more information becomes available it is easier to adapt and refine the techniques. If a following distance of 0.1 s is assumed for increased lane capacity, there must also be a discussion of the assumption on how lane changes will be managed. What are the assumption being made for speed: will all AVs be traveling at the same speed and only be able to slow down to find an acceptable gap for a lane change?

Another step is to avoid averages and oversimplifications of uncertainty. Rather than discussing the average latent demand in an entire network, perhaps the demand can be looked at on highways, arterials, or streets in the central business district. Is that only during the peak period or daily traffic? Oversimplification can lead to generalizations that are unhelpful at best and misleading at worst.

Third, planners must also utilize best practices for evaluating alternatives in the face of uncertainty. Decision analysis has many methods to evaluate uncertainty and assess system value.

Finally, planners should continue to reduce the uncertainty by any possible means. Bringing OEMs to the conversation to provide expected operating specifications, even if not the technical specifications for how they expect to achieve them, could be valuable. Planners ought to look at all of the pilot programs and work in tandem with them, helping to design tests and data collection so as to be as useful as possible for future research and planning.

Our transportation plans and our understanding of CAVs will continue to evolve. With diligent research, an ever-expanding tool set, and a vision of the future possible through automation and connectivity, we have the ability to harness these technologies to shape our transportation networks to improve our cities and the quality of life of those who live there.

References

1. *Federal Autonomated Vehicle Policy*. National Highway Traffic Safety Administration. 2016.
2. Stanford.edu. *SAE Levels of Driving Automation*. Stanford Law School. 2016. http://cyberlaw.stanford.edu/files/blogimages/LevelsofDrivingAutomation.pdf. Accessed July 24, 2017
3. Federal Communication Commission. *FCC 03-324*. 2003.
4. Sens, T. Self driving cars: enroute to 5G. *Nokia*. June 23, 2016. https://blog.networks.nokia.com/mobile-networks/2016/06/23/self-driving-cars-enroute-5g/. Accessed July 2, 2017.
5. Federal Highway Administration. MPO Boundary Data. https://hepgis.fhwa.dot.gov/fhwagis/. Accessed June 5, 2017.
6. DuPuis, N., M. Cooper, and B. Rainwater. *City of the Future Technology and Mobility*. National League of Cities. 2015. Accessed June 15, 2017.
7. *Long Range Transportation Plan: Outlook 2035 Because Mobility Matters*. Orange County Transportation Authority (OCTA). 2014.
8. *Maximize 2040: A performance based Transportation Plan*. Baltimore Regional Transportation Board (BRTB). 2016.
9. *Impact of Future Technology in the 2045 LRTP*. Miami-Dade Transportation Planning Organization (TPO). Prepared by The Corradino Group. 2017.
10. *Intelligent Transportation Systems Master Plan*. MetroPlan Orlando. 2017.
11. *Traffic Safety Facts 2008*. National Highway Traffic Safety Administration. U.S. Department of Transportation. Washington, DC. 2008. DOT HS 811 170.
12. HCM 2000: *Highway Capacity Manual*. Transportation Research Board. 2000.
13. *Traffic Congestion and Reliability Trends and Advanced Strategies for Congestion Mitigation*. Cambridge Systematics. Federal Highway Administration, 2005.
14. Farmer, D. L. Autonomous Vehicles: the Implications on Urban Transportation and Traffic Flow Theory. *ITE Journal*, Vol. 86., 2016.
15. Shladover, S., D. Su, and X. Lu. Impacts of Cooperative Adaptive Cruise Control on Freeway Traffic Flow. *Transportation Research Record: Journal of the Transportation Research Board*, 2012. 2342: 63–70.

16. Talebpour, A., and H. S. Mahamassani. Influence of Connected and Autonomous Vehicles on Traffic flow Stability and Throughput. *Transportation Research Part C*, Vol. 71, 2016, pp. 143–163.
17 Department for Transport. *Research on the Impacts of Connected and Autonomous Vehicles (CAVs) on Traffic Flow.* Summary Report. 5145311. 2016.
18. *Proposal for Automated Vehicle Proving Grounds Designation.* Texas AV Proving Ground Partnership, 2016.
19. Fagnant, D. J., and K. Kockelman. Preparing a Nation for Autonomous Vehicles: Opportunities Barriers and Policy Recommendations. *Transportation Research Part A: Policy and Practice*, Vol. 77, 2015, pp. 167–181.
20. Pascale, J. Gov. McAuliffe signs bill to allow autonomous delivery robots — but it's unclear when they launch in Virginia. *Pilotonline*. June 1, 2017. https://pilotonline.com/news/government/politics/gov-mcauliffe-signs-bill-to-allow-autonomous-delivery-robots-but/article_944e8f12-5f75-5e86-bde7-10b0979c55ac.html. Accessed July 18, 2017
21. Krechmer, D., J. Osborne, J. Bittner, M. Jensen, and E. Flanigan. *Connected Vehicle Impacts on Transportation Plannnig Technical Memorandum #2: Connected Vehicle Planning Processes and Products and Stakeholder Roles and Responsibilities*. FHWA-JPO-15-246. FHWA, U.S. Department of Transportation, 2015.
22. Grant, M., J. D'Ignazio, A. Bond, and A. Mckeeman. *Performance Based Planning and Programming Guidebook*. FHWA-HEP-13-041. FHWA, U.S. Department of Transportation, 2013.
23. Parnell, G. S. *Decision Analysis in One Chart*. Decision Line, Newsletter of the Decision Sciences Institute. May 2009.
24. Parnell, G. S., T. Bresnick, S. Tani, and E. Johnson. *Handbook of Decision Analysis*. Wiley & Sons, 2013.
25. Hawkins, A. J. Uber's self-driving cars are now picking up passengers in Arizona. *The Verge*. Feb 21, 2017. https://www.theverge.com/2017/2/21/14687346/uber-self-driving-car-arizona-pilot-ducey-california. Accessed July 24, 2017.
26. Enwemeka, Z. Wbur Bostonomix. *Boston Allows Self-Driving Car Pilot Program to Expand to More Roadways*. April 25, 2017. http://www.wbur.org/bostonomix/2017/04/25/boston-self-driving-car-pilot-expansion. Accessed July 10, 2017.
27. DOT. U.S. Department of Transportation Designates 10 Automated Vehicle Proving Grounds to Encourage Testing of New Technologies. *Transportation.gov-briefing room*. January 19, 2017. https://www.transportation.gov/briefing-room/dot1717. Accessed July 20, 2017.
28. Texas AV Proving Ground Partnership. Texas AV Proving Ground Partnership Proposal. *Texas A&M University*. December 19, 2016. https://tti.tamu.edu/group/connected-vehicles/files/2014/02/Texas-AV-Proving-Ground-Partnership-Proposal-12192016-Final-updated.pdf. Accessed July 3, 2017.
29. Wisconsin Automated Vehicle Proving Grounds. Automated Vehicles. *Wisconsin Automated Vehicle Proving Grounds*. 2017. http://wiscav.org/. Accessed July 2, 2017.
30. *Regional Transportation Plan 2045 Maintaing the Vision for a Sustainable Region*. New York Metropolitan Transportation Council (NYMTC). 2017.
31. City of Columbus. Smart Columbus Projects. *The City of Columbus US*. 2017. https://www.columbus.gov/smartcolumbus/projects/. Accessed July 2, 2017.

The Standing Committee on Statewide Multimodal Transportation Planning (ADA10) peer-reviewed this paper (18-06654).

Article

TRR
JOURNAL OF THE TRANSPORTATION RESEARCH BOARD

Transportation Research Record
2018, Vol. 2672(51) 20–30

DOI: 10.1177/0361198118792767
journals.sagepub.com/home/trr
SAGE

A Framework for Identifying Recurring Bottlenecks in the State Planning Process in Colorado

Oana Deselnicu[1] and Jason Wallis[1]

Abstract

Traditional methods used in transportation planning to identify bottlenecks and mobility issues, such as the Volume-to-Capacity Ratio and the Planning Time Index, have limited usefulness in identifying the exact location and extent of bottlenecks. Moreover, existing bottleneck identification tools flag all bottleneck types without distinction, despite the fact that strategies and resources used to address each type are different. This research first proposes a taxonomy of congestion to distinguish between different types of bottlenecks. It then describes a new methodology for identifying the location of recurring bottlenecks. Recurring bottleneck locations must experience a reduction in speed, an upstream accumulation of vehicles, and recurrence at the same location over three consecutive months. The methodology is currently used in long-term transportation planning and project selection in Colorado to identify and address the most severe bottleneck locations in the state.

In 2015, congestion on Denver regional interstates and major regional and principal arterials alone was estimated to cost $1.3 billion, a number projected to increase to $3.3 billion by 2040 according to the Denver Regional Council of Governments' 2015 Annual Congestion Report (*1*). Bottlenecks cause travel delays, contribute to traffic incidents, interfere with commercial deliveries and pick-ups, and worsen air quality. The negative impacts of congestion include lowered productivity and increased costs for both personal and commercial vehicles—affecting economic growth—as well as adverse effects on quality of life. While bottlenecks are not the only cause of congestion, they are certainly a contributing factor to general congestion (*2*).

Transportation agencies actively address congestion using operational, technological, and enhanced infrastructure strategies, among others. The intention of this methodology is to enhance the analysis used in transportation planning to assess the current environment, identify trends, and provide strategies to mitigate, monitor, or address congestion on the transportation system.

As part of long-range, regional and statewide planning, the Colorado Department of Transportation has developed a methodology to identify bottlenecks in the state that can be addressed in a multi-year plan. The new methodology arose from a need to balance long-term strategies and investments with temporary causes of congestion, such as weather, incidents, or event traffic which are usually addressed through transportation management and operations strategies. However, recurring congestion often requires long-term strategies and investments. To fill this gap, a data-focused approach is needed to inform these strategies and serve as a resource for those making funding decisions. Congestion events most suitable to be included in long-term planning are generally long-standing issues that require concentrated resources such as time, capital improvements, and funding. Including these types of congestion in the planning process can create an avenue for resource allocation and prioritization to the most severe locations. While existing resources and analyses on this topic frequently make a distinction between recurring and non-recurring congestion events, they do not offer distinct methodologies to identify each. This research intends to better define the elements of congestion, to propose a methodology for identifying recurring, localized congestion, and mitigation strategies during the planning process.

[1]Colorado Department of Transportation, Denver, CO

Corresponding Author:
Address corresponding to Oana Deselnicu: oana.deselnicu@state.co.us

Objectives

This research has three primary objectives. First, this research proposes a classification of congestion. While congestion typologies similar to the one developed in this paper have been proposed in the past, the methods to identify each form of congestion are not clearly defined. As such, the second objective of this research is to develop a methodology for identifying recurring congestion that occurs at specific locations. Finally, the third objective is to illustrate how these results are used to enhance the transportation planning process.

Terminology

The terminology and definitions for congestion and its elements, the components of a bottleneck, and other terms used throughout this document research may have different meanings depending on their context or application. The following definitions outline the context for our subsequent analysis.

Congestion is a loosely defined term broadly used to describe a range of conditions and causes leading to a less efficient transportation system. In this research, the term congestion is used in the broad sense; see Figure 1. Congestion can be geographically generalized or localized. Generalized congestion is observed across an entire corridor, is widespread in nature and cannot be pinpointed to a specific location, overwhelming that entire corridor. Localized congestion can be identified at a particular location and is more granular in nature than generalized congestion. Both congestion types can be recurring or non-recurring. Recurring congestion has a predictable pattern, usually temporal, and can be considered status quo for the system. Non-recurring congestion is typically caused by a factor that is not constant in the system. The purpose of this analysis is to find recurring localized conditions that contribute to localized congestion or bottlenecks.

A *bottleneck* is an equally vague and difficult to standardize concept subject to change in relation to the purpose of the analysis performed. Some definitions are very broad: a bottleneck is a "hindrance, impediment or constriction of traffic flow" (*3*, *4*). Other definitions are more specific: bottlenecks are locations experiencing an upstream queue and free-flowing downstream traffic (*5*).

Based on the congestion typology above, the definition of a bottleneck in this research is a *recurring* reduction in speed at a specific location and time of day, accompanied by a *reduction in speed* and queue forming upstream and free-flowing traffic downstream. Each bottleneck displays a *head-and-tail pattern*, where the head is the *bottleneck*, the forward-most point in the direction of travel where the first reduction in speed is registered. A *tail* is the build-up of vehicles traveling at reduced speeds upstream and directly adjacent to the bottleneck. The intent of this methodology is to precisely identify the bottleneck head, where the problem begins, as a way to relieve traffic behind it.

Related Works

Most of the literature related to bottleneck identification acknowledges that there is recurring and non-recurring congestion with distinct characteristics, but does not always customize the analysis to each specific type of

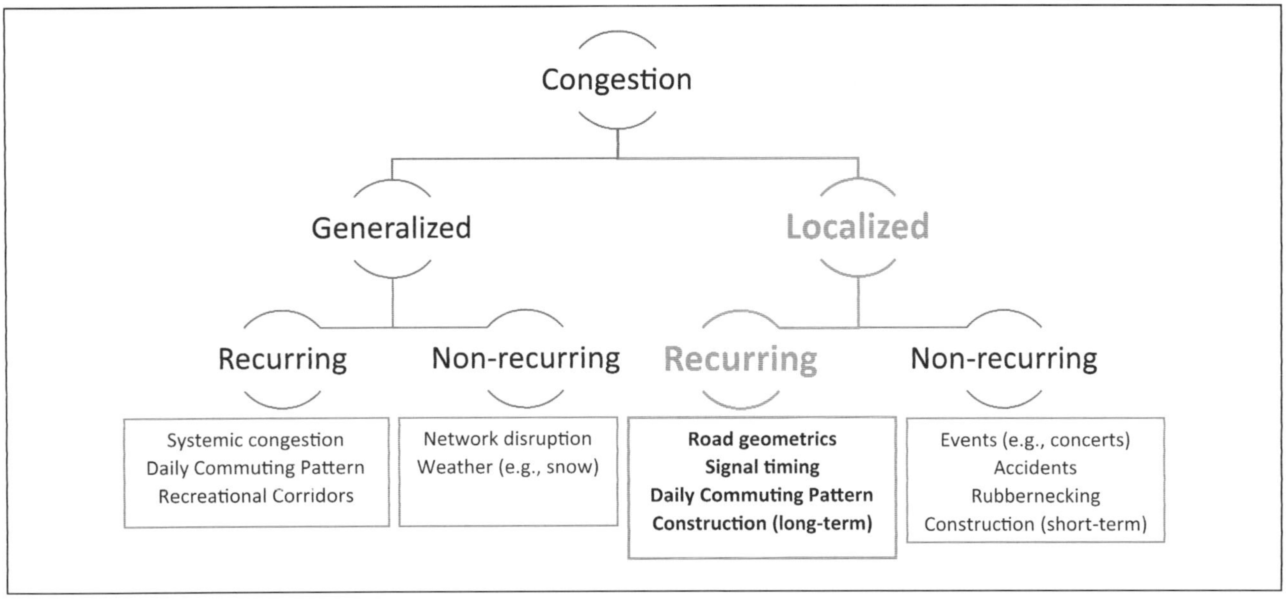

Figure 1. Elements of congestion.

congestion. Recurring localized congestion, or bottlenecks in this analysis, has a specific location and can be tracked over time. Non-recurring congestion, be it localized or system-wide, is temporary, and may not be tracked over time.

A recent Transportation Research Board (TRB) paper using probe data to identify bottlenecks calculates an impact value for each bottleneck as the sum of lengths of every road segment included in the bottleneck occurrence (*6*). Locations with a higher impact value represent recurring congestion while locations with a lower value are non-recurring congestion. While the paper is very successful in proposing a new methodology for tracking bottleneck changes dynamically and thus improving bottleneck location accuracy, it is not clear that the distinction between recurring and non-recurring bottlenecks identified by the impact value is sufficient.

Other studies proposing a methodology for recurring bottleneck identification use a static congestion speed threshold, such as 40 mph, to find bottlenecks (*2*, *7*) or use data sources such as loop detectors (*7*). A decrease in speed alone, especially a static one expressed in miles per hour as opposed to a percent decrease, may not uncover recurring bottlenecks which operate on a few different dimensions, such as speed, time of day, month, year, and location in space (for recurrence and queuing behavior). The same can be said about measures of congestion such as Travel Time Index and Planning Time Index, or Hours of Delay (*8*) that take a snapshot of current conditions but need to be tracked over time to identify recurring problems.

The 2017 Wisconsin State Freight Plan explains the difference between recurring and nonrecurring bottlenecks very well, but does not employ customized methodologies to identify the two categories individually. A bottleneck is identified if travel speed drops below 60% of free-flow speed for 15 minutes or longer and ends with an uninterrupted period of 15 minutes above the threshold speed (*9*). Both recurring and nonrecurring bottlenecks are identified this way.

The available literature on bottleneck identification contains all of the elements for identifying recurring bottlenecks, such as speed reduction, queueing, and recurring behavior over time, but we argue that there is no current methodology that uses them all simultaneously. The proposed approach attempts to combine all of these elements to identify recurring bottlenecks that can be addressed in a long-term planning process.

Data

This approach is based on speed data provided by INRIX. INRIX is a private data provider that integrates a large variety of sources (such as GPS-based probe vehicles and mobile devices) to calculate real-time speed on the roadways (*2*). Two measures are used from the INRIX dataset: speed and reference speed. Speed is the average real-time speed experienced on the roadway. Reference speed is defined by INRIX as an uncongested "free-flow" speed determined for each road segment using archived data. This is used as the baseline for non-congested speed.

We obtained Historical speed and reference speed data for a subset of Colorado State Highways, amounting to 4,156 centerline miles and 12,116 total lane miles, or roughly 45% of the state system. This includes the entirety of the state's interstate highways, expressways in urbanized areas and principal arterials in rural areas, and other routes important for regional connectivity.

Speed data for 2014, 2015, and 2016 were aggregated for each road segment or traffic messaging channel (TMC) for each month of each year. This resulted in a monthly matrix of ninety-six 15-minute epochs on the horizontal axis and the list of analyzed road segments on the vertical axis, populated with the average speeds for each segment at each time interval. Epochs are 15-minute time intervals in a 24-hour period, and represent the time units within which potential bottleneck locations are evaluated.

Because bottlenecks are identified for each corridor and direction of travel, this data preparation process was completed separately for each corridor, direction of travel, and month of the year. TMCs, referred to as road segments throughout this paper, serve as the basis of the road network segmentation in this analysis. Although TMCs are used in this research, they are not the only road segmentation that can be used. Any segmentation defined by a user, provided it is granular enough, can be used, although we do not recommend road segments longer than a few miles. Increased granularity of road segments is helpful in accurately identifying bottlenecks. Longer road segments can experience a range of speeds with high variance and may falsely identify bottlenecks. Moreover, pinpointing the cause and location of a bottleneck is likely to be less accurate on long roadway segments.

While the INRIX speed dataset is one of the most complete, extensive, and accurate speed datasets available to state DOTs, it is not without flaws. Sometimes, segments may not have enough real time samples. In that case, INRIX substitutes those records with the historical average and the reference speed, in this order. Rural corridors and smaller roads suffer from this disproportionately. A 2016 study comparing field data to INRIX real time speed data in Colorado found the accuracy and reliability of INRIX data to meet Federal Highway Administration (FHWA) standards on interstates, but not on lower volume roads (*10*). Validating INRIX data

Table 1. Example of Monthly Data of Percent Speeds below Threshold (%)

Epoch Segment	00:00	00:15	...	...	...	23:45
Segment 1	2.3%	0%	...	...	...	0%
Segment 2	0%	7.6%	...	...	...	12%
...	...	...	...	...	...	...
Segment *n*	9%	0%	...	...	...	4.7%

is often difficult, because it is expensive and time consuming for a DOT to collect accurate field data for comparison, and not all locations can be sampled. INRIX historical data could not be assessed in this study (*10*). While corridor validations may need to be performed before specific results on rural corridors are used, for methodology development and testing we found the data quality and coverage from INRIX to be sufficient for refining the overall process described in this paper. To verify individual results, we performed anecdotal validation with local experts.

Methodology

Our approach determines a bottleneck location based on three conditions: a reduction in speed, queueing upstream and free-flow speeds downstream of the bottleneck, and recurrent behavior during the same time-interval for three consecutive months. Each condition is discussed in detail below.

Speed Reduction

Speed-based bottleneck identification specifies a speed threshold below which the road segment is considered congested. After testing several speed reduction thresholds we determined that 10% below reference speed was appropriate for the data. This is referred to as "threshold" throughout the paper. The roadways analyzed with this approach are located across the state in both urban and rural environments, and the threshold needed to reconcile both environments. Speed thresholds higher than 10% were tested and it was found that their increased restrictiveness eliminated some of the rural bottlenecks. At the same time, thresholds below 10% are too inclusive thus diminishing the usefulness of this approach in comparison to traditional congestion identification methods. This speed threshold is intended to be adjusted based on corridor environment (rural or urban) and other specifics of the investigated sample.

The matrix of time of day and road segments is recalculated to display the percent speed below threshold for each segment at each time interval for each month (see Table 1). When speed is within 90% of reference speed, there is no delay, meaning the percentage of speed reduction is zero (0%). Any speeds within or above the threshold are also coded as zero (meaning no delay). Any speeds where the reference speed is less than or equal to 30 mph are coded also as zero percent (meaning no delay). Due to the characteristics of the selected network, low reference speeds were often indicative of "Main Street" segments of towns on rural highways, segments we did not want to flag as bottlenecks.

Plotting all speed reductions (as percent) for a corridor over all of the months of an entire year offers a visual indication of which months and parts of the corridor experience the highest speed reductions (see Table 1). For example, at 00:15, Segment 2 in Table 1 experienced a 7.6% decrease in speed from the established speed threshold.

A visual interpretation of Table 1 above is provided in the heat map in Figure 2. It displays the epochs on the horizontal axis and the road segments in the direction of travel on the vertical axis. The color variation shows the percent reduction in speed experienced by each road segment at each time frame. While for the locations experiencing reduced speeds, the time of day, and the length of the tails are evident in this graphic, it is not clear which locations experience localized versus generalized congestion, and in particular, recurring bottlenecks.

Head-and-Tail Pattern

Another condition of recurring bottlenecks is the vehicle queue (tail) traveling at reduced speed upstream of the bottleneck while free-flow speeds are experienced downstream of the bottleneck. Our algorithm specifically looks for this head-and-tail pattern by scanning each road segment in the direction of travel for each 15-minute interval. If, for the same epoch, a road segment experiences a reduction in speed under threshold AND has at least one other directly adjoining upstream segment also experiencing a speed reduction, while the adjoining segment directly downstream experiences free flow, then the location at that time interval is flagged as a potential bottleneck for that month. The process is repeated for each month of a year, resulting in twelve lists of flagged locations, one for each month. The flagged locations

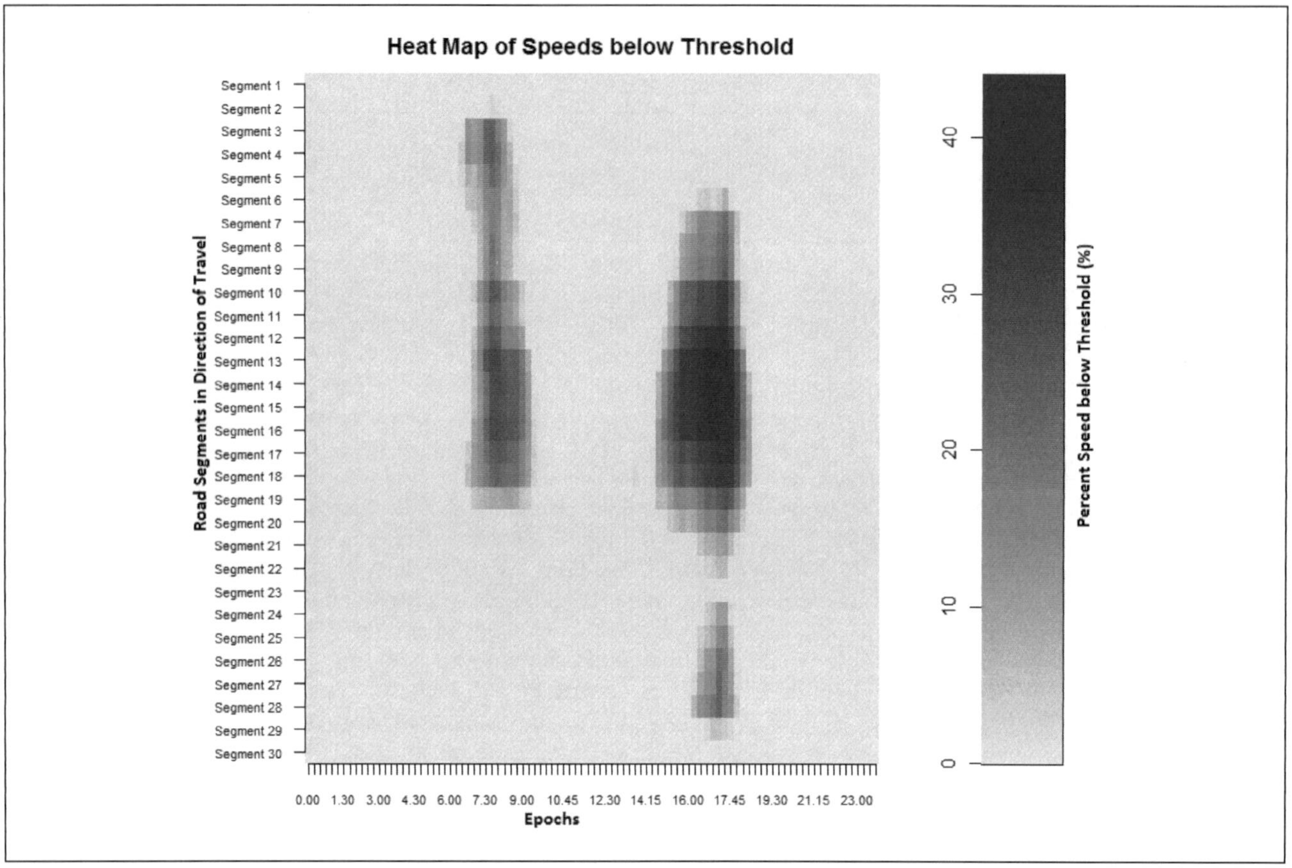

Figure 2. Percent speed below threshold on I-70 WB (Chambers Rd. to Federal Blvd.).

represent potential bottlenecks. While the algorithm ensures that there is a tail for each flagged location, bottleneck tails are not recorded in the results. Figure 3 below represents the same corridor as the Figure 2 heat map but only displays bottleneck heads. Locations with tails upstream and free-flowing speeds downstream are identified and flagged in this graphic. Visually, the heat map is already greatly simplified by displaying only the problematic locations for each segment and time interval.

Recurrence

The last condition applied for identifying recurring bottlenecks is a temporal one. If the bottleneck re-occurs due to consistent causes, we are able to consistently flag it across several months of an entire year. The algorithm captures this behavior by checking every flagged bottleneck location identified in the previous step at each time interval each month to see if it has also occurred in the past two previous months. For January, for example, it checks the months of November and December of the previous year at the same location and time interval. For February, it checks January of the current year and December of the previous year, and so on.

Figure 4 shows the average percent speeds lower than threshold on the vertical axis for three months on one corridor. The speed reduction for a month is averaged across all epochs in that month for graphing purposes. The horizontal axis displays the road segments on the corridor in the direction of travel. Visually, it is apparent that Peak A located around Road Segment 7 experiences a reduction in speed across all three months. January, February, and March experience speed reductions of 5.21%, 3.6%, and 2.39%, respectively. Conversely, Peak B, located around Road Segment 70, only experiences a speed reduction during January (3.88%) and February (4.39%) and no delay in March (0%). Since three consecutive months of recurrence are required for a long-term bottleneck, Peak A is flagged as a bottleneck for the month of March, whereas Peak B is not flagged as a bottleneck.

We tested different levels of recurrence (e.g., two months and three months) representing varying levels of restrictiveness and decided three consecutive months is a suitable period for measuring recurrence in our sample.

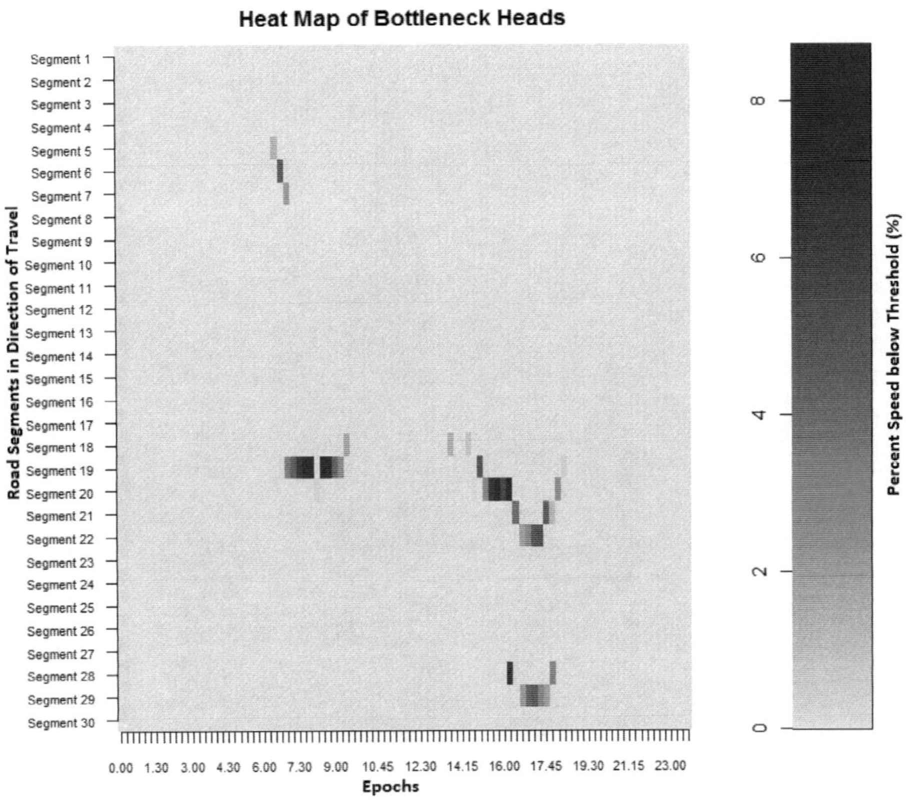

Figure 3. Locations meeting criteria 1 and 2 are flagged as bottlenecks on I-70 WB (Chambers Rd. to Federal Blvd.).

Tests show that two-month recurrences still capture most construction projects and other temporary events, while three-month recurrences eliminated all but the most severe of them.

A location is marked as a recurring bottleneck if all three of the following conditions are met: 1) a reduction in speed of at least 10% of reference speed; 2) a head-and-tail pattern with vehicles queueing upstream and free-flow speeds downstream; and 3) recurrence at the same location and epoch across three consecutive months.

Figure 5 is the last in the series of heat maps illustrating the gradual progression of this approach. Previously flagged locations that do not experience a three-month occurrence are eliminated, providing a list of locations that meet all of the previously outlined criteria. Each analyzed corridor is provided such a list for each direction of travel for each month.

For a final list of a corridor's bottleneck locations over the course of a year, all twelve monthly lists are combined. A location can be identified in more than one monthly list. The final yearly list counts how many months out of the year each location occurs, providing a sense of the relative severity of each location.

Discussion of Results

The results of this methodology are useful in several ways. First, this approach generates a yearly list of recurring bottlenecks that can be tracked from year to year to eliminate inconsistencies, establish severity, and conduct prioritization. The list of bottlenecks for each year is identified using the three requirements defined above. When the analysis is repeated every year and lists of bottlenecks are available for several years in a row, they can be compared with each other to see which locations endure over time. Tracking locations over time allows users to focus on the most severe recurring locations while eliminating long-term construction, other events, and data issues. Year-to-year comparisons are not a requirement of this methodology, but rather a way to curate a list of bottlenecks if the analysis is repeated every year.

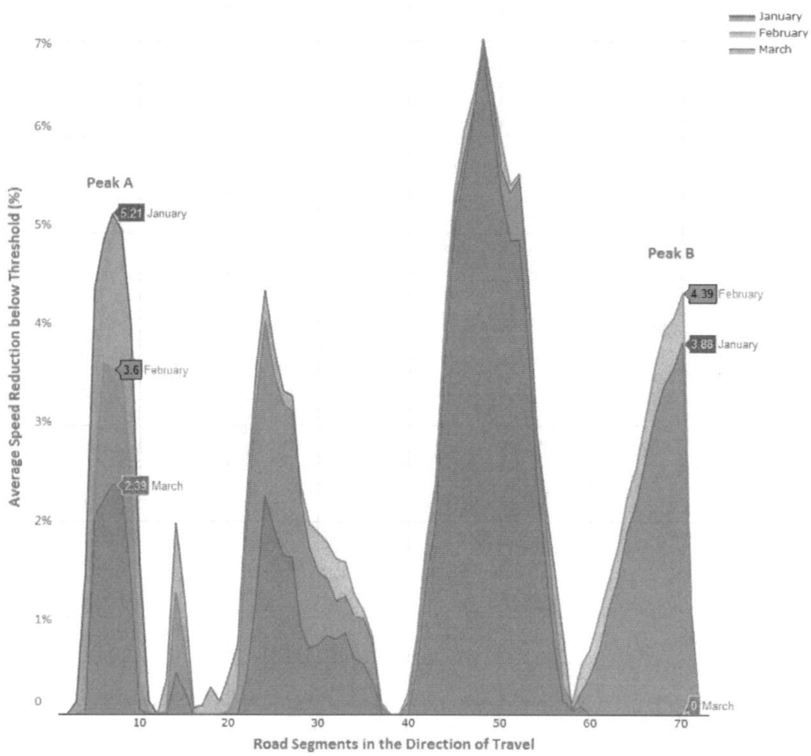

Figure 4. Percent speed reduction recurring monthly along I-25 SB (CR 34 to Briargate Parkway).

For example, to finalize the list of 2016 bottlenecks in our roadway sample, only identical locations from consecutive years are kept, in addition to a handful of locations flagged in only 2015 but having recurred many months out of 2015 (see Figure 6). However, most severe locations tend to be consistently identified from year to year.

As 2017 data become available, the 2016 list from this analysis will go through a similar process to arrive at the 2017 bottleneck locations. The most recent list of bottlenecks is intended to be a dynamic document that reflects the best information available.

One example of a bottleneck that can be eliminated through a year-over-year comparison is the long-term bridge replacement construction project on I-70 and Havana in Denver. The project, lasting until the end of 2016, began at the intersection of I-70 WB and Havana in May 2015, causing intermittent lane closures at that location. Since this is a recurring event for all of 2016, it was picked up as a bottleneck in 2016 (see Figure 7) while the location was not flagged in the 2015 data. While 2017 data are not yet available, we hypothesize that Havana and I-70 will not be a bottleneck in that dataset either. By overlapping results from different years, this long-lasting construction event will ultimately avoid classification as a bottleneck. This allows the most severe recurring locations to be prioritized while long-term construction and other similar temporary events are eliminated.

Second, the new methodology provides more granular results in comparison with traditional methods. The proposed approach identifies a specific location where the problem begins for each direction of travel. When compared with traditional congestion measures such as the volume-to-capacity (V/C) ratio, this methodology yields similar—but more granular and useful—results. The V/C ratio may show an entire corridor as congested (see Figure 8, first panel). This result is not very useful if a transportation agency hopes to make specific and focused investment decisions or to prioritize projects. When heads and tails are mapped together in the proposed methodology, the resulting pattern matches the V/C ratio quite well, though not identically (Figure 8, second panel). The difference in coverage results from the V/C ratio compounding both directions of travel, while our methodology shows results for each direction separately. Figure 8 (panels 2 and 3) displays only northbound travel, but the coverage can be reconciled when both directions (northbound and southbound) are

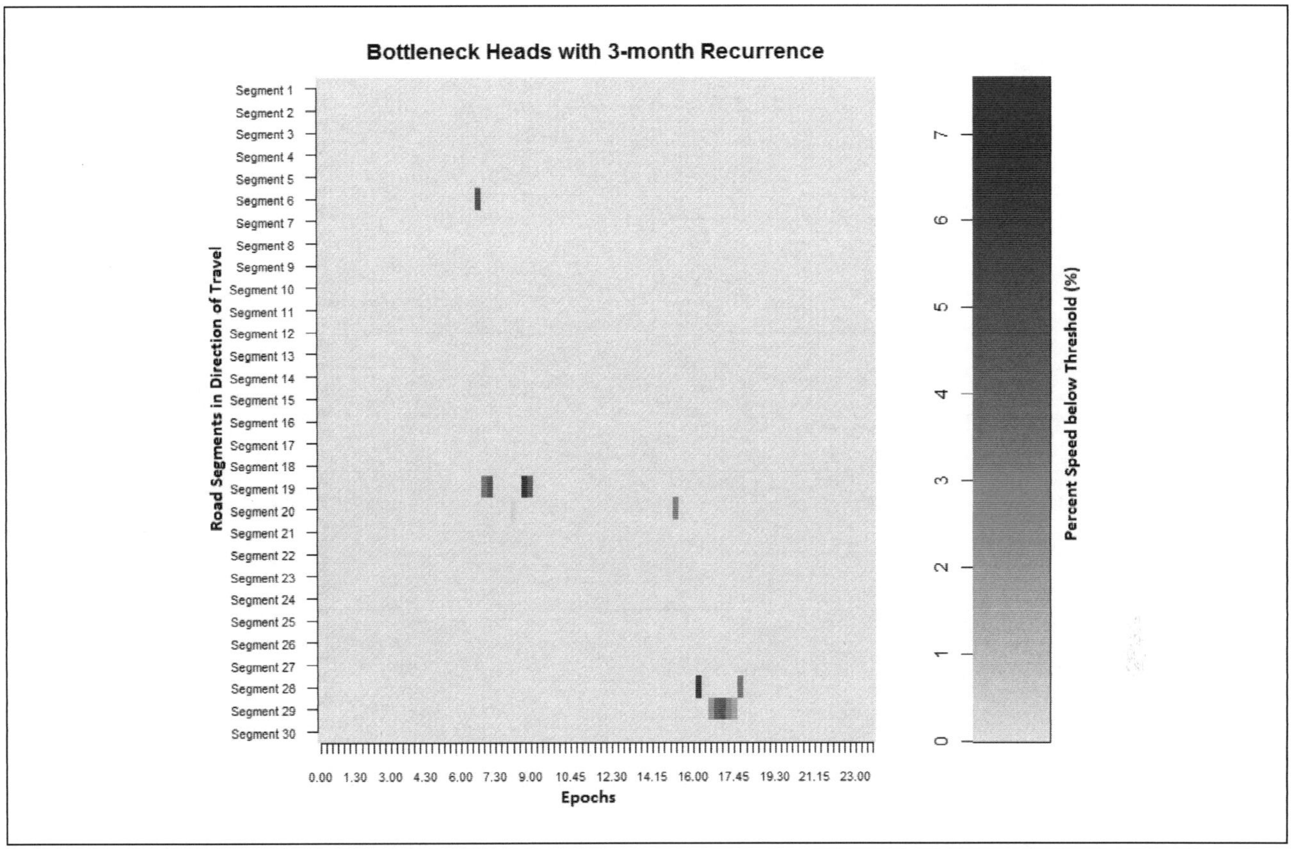

Figure 5. Locations meeting all criteria are flagged as bottlenecks on I-70 WB (Chambers Rd. to Federal Blvd.).

displayed. To illustrate the power of the new approach, when tails are removed in the third panel, the bottleneck heads pinpoint the causal locations on that facility. Not only can results be disaggregated by direction of travel with the new approach, but specific bottlenecks can be identified enabling resources to be focused at the appropriate specific locations.

Third, this methodology identifies locations of urban-like congestion (such as signal timing) in a rural setting, where the V/C ratio or other tools are usually too blunt to distinguish problematic locations. For example, we identified a bottleneck location on US 34 in a rural setting (northeast Colorado). This is a rural section of US 34 where congestion is rare but bottlenecks still exist due to traffic signal timing. In this example, the analysis identified bottlenecks in both directions of travel at the intersection of US 34 and CR 17 (Figure 9). Other tools are not available to pinpoint specific problematic locations in rural settings. In this scenario, the bottleneck can be addressed through operational strategies and signal timing.

By increasing location granularity and using the latest available data to dynamically identify localized recurring congestion issues, the approach proposed in the paper can be a valuable tool for planners and transportation agencies to help allocate congestion mitigation resources to their most productive uses over the long run.

In conclusion, this paper builds on previous literature to first propose a taxonomy of congestion types and then describes a methodology to identify recurring localized congestion for transportation planning purposes. While various types of congestion have been acknowledged in other studies, the methods to identify each specific type have been slower to evolve. Usually, bottleneck identification methods do not make a distinction between recurring and non-recurring bottlenecks. This paper aims to make this methodological distinction. The novelty lies in a series of conditions that distinguish between recurring and non-recurring bottlenecks and in the ability to pinpoint the location where bottlenecks start to form. This is very useful to practitioners and planners interested in identifying these specific issues and using the correct resources to alleviate them.

Colorado Use Cases

In Colorado, this methodology is used for several different purposes. First, to identify freight bottlenecks

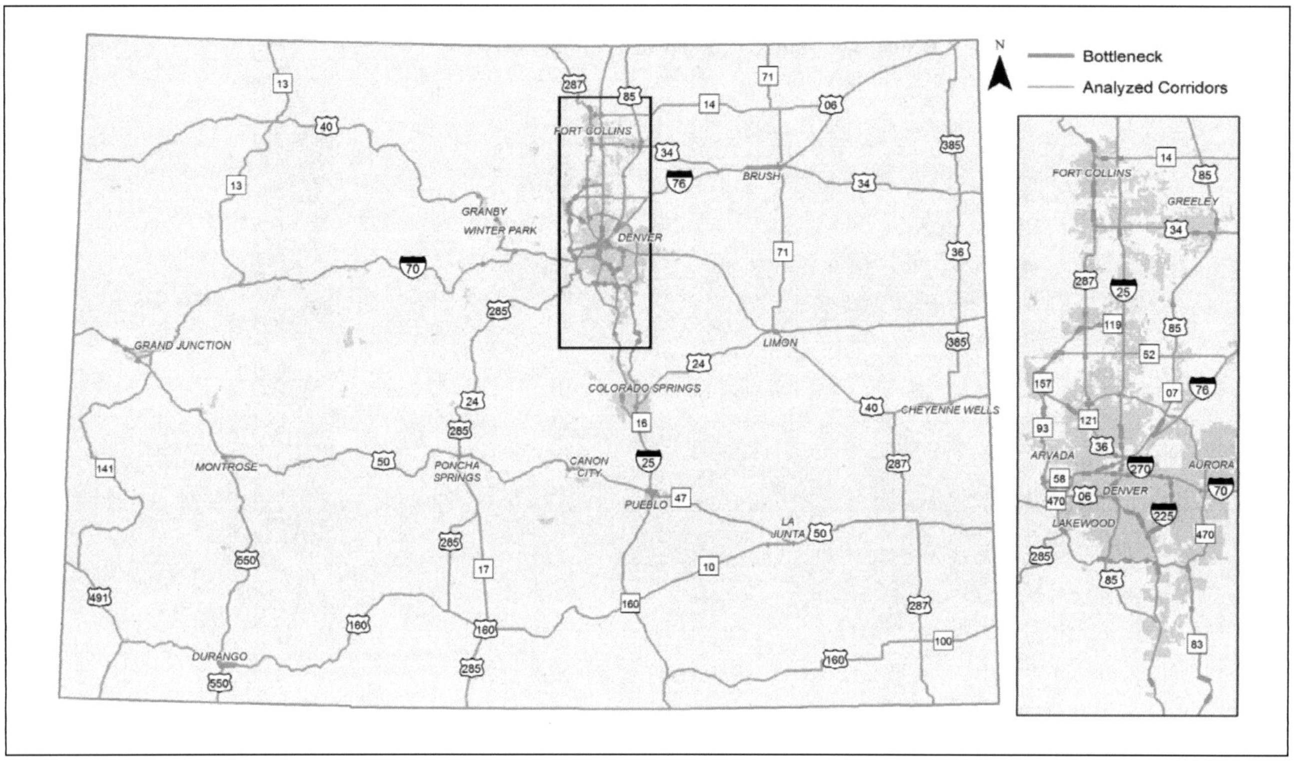

Figure 6. Final list of 2016 bottlenecks.

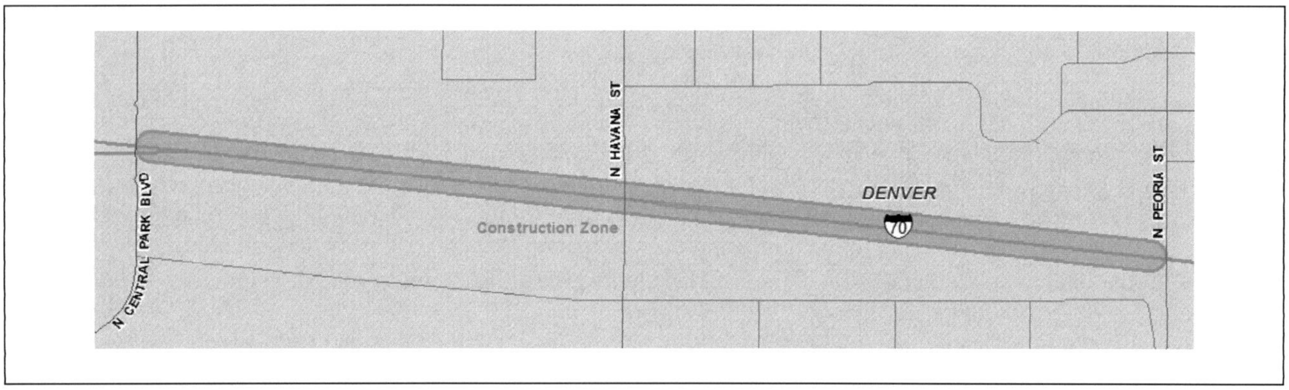

Figure 7. I-70 WB and Havana Bridge replacement.

for the Colorado Freight Plan. After identifying freight bottlenecks for the Colorado Freight Plan, this list will be updated yearly with new bottlenecks that appear in the system and remove ones that are no longer issues.

Second, these bottlenecks will also be used as one of the criteria to evaluate projects for National Highway Freight Program funds. With unlimited funding needs but limited funds, projects are evaluated in an increasingly competitive process to allocate funds to the highest-benefit projects. For example, if a freight project proposes mobility improvements and is located in an identified freight bottleneck, it may receive a higher score and have a better chance for funding.

Another area where this methodology will be used is the Colorado 2045 Statewide Plan. For this application, the methodology will be extended to the entire state, provided adequate underlying data quality. The identified bottlenecks will be included in a competitive project selection process where their bottleneck status may help in securing funds.

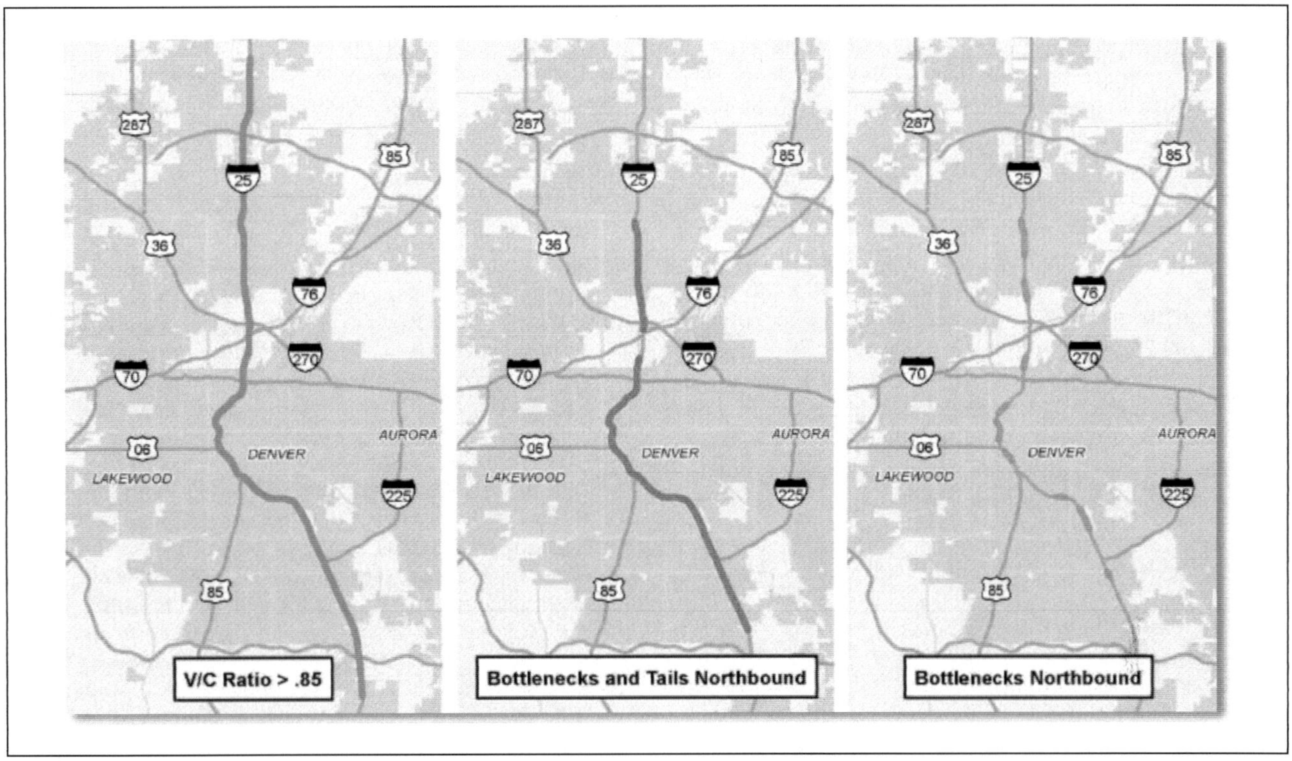

Figure 8. Comparison of V/C ratio, bottlenecks and tails, and bottlenecks on I-25 NB in the Denver area.

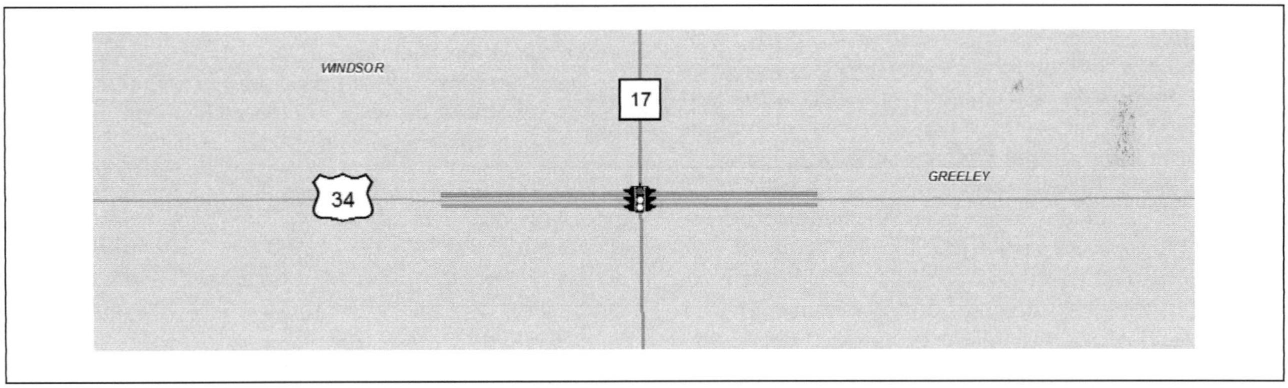

Figure 9. Signalized intersection bottlenecks on rural US34 and CR17 (EB and WB).

Limitations and Future Work

The approach outlined above is an evolving process that will mature with use to serve the transportation planning purposes of Colorado. There are still several issues that require further exploration and analysis. For example, in the current approach, speed data for all weekdays and weekends is included together in the analysis. However, analyzing weekends and holidays separately may be important for some facilities. The Intersatate-70 West corridor in Colorado, for instance, tends to be a recreation corridor and may behave very differently on weekends, seasons, and holidays.

Another limitation for further exploration is the different behaviors of rural and urban facilities. The speed threshold requires adjustment based on corridor location. The speed threshold can change the resulting bottleneck locations and should be adjusted carefully based on the characteristics of the investigated corridors. Rural corridors tend to have longer road segments that can interfere with bottleneck identification accuracy (the longest in our sample was 52 miles). Longer segments experience a larger distribution of speeds which can distort the average speed recorded for that segment. Adopting a more granular segmentation, such as the XD instead of the TMC network in INRIX, must be tested as a potential improvement for the accuracy of results.

While we experienced no issues in this analysis with missing or incomplete data, data quality is a further potential issue to be considered in other samples. A data validation process to ensure each road segment has an acceptable random sampling error should also be included if results are used for project selection or decision-making (*2*).

The benchmark against which congestion is measured is another item that may be tested in the future. Various studies employ different measures ranging from a static posted speed to dynamic free-flow speed (defined in various ways), to a top percentile of speed, to reference speed (used in this analysis), and to several others. There is no consensus currently as to which benchmarking measure is superior.

Identifying the cause of the bottlenecks under the current framework is another limitation. Currently, this needs to be conducted following this analysis because the dataset does not contain the relevant information (such as road geometrics, or the presence of a signal) to automatically identify the bottleneck cause. If conflation can be done between the speed dataset and road geometrics information collected by the state, an automated algorithm can be created to check if bottleneck locations coincide with steep grades, curves, lane drops, or other issues. Once a cause is identified, prioritizing bottlenecks to be addressed within a fiscally constrained budget is the next logical step. Statistics on the length of the bottleneck, the number of months it occurs in a year, its average duration, and traffic counts (which can be folded into a vehicle-hours of delay measure), are good metrics to include in a bottleneck prioritization process. In Colorado, the intent is to use the prioritized bottlenecks as part of a scoring system to prioritize investment in our infrastructure by ranking projects competing for funding.

Author Contributions

The authors confirm contribution to the paper as follows: study conception and design: the study concept resulted from a legitimate research activity encountered by the authors. Data collection: authors have not collected the data themselves, but are subscribed to the RITIS/INRIX service that collects and distributes the data. Analysis and interpretation of results: authors confirm that the data processing algorithm is original and take responsibility for results and their interpretation within the paper. Draft manuscript preparation: authors have prepared the manuscript themselves. All authors reviewed the results and approved the final version of the manuscript.

References

1. Denver Regional Council of Governments. *2015 Annual Report on Traffic Congestion in the Denver Region*. 2016. https://drcog.org/sites/drcog/files/resources/2015%20Annual%20Traffic%20Congestion%20Report_0.pdf.
2. Florida Department of Transportation. *Strategic Intermodal System (SIS) Bottleneck Study, Technical Memorandum No 2, Methodology to Identify Bottlenecks*. Florida Department of Transportation, Tallahassee, FL, 2012. http://www.fdot.gov/planning/systems/programs/mspi/pdf/Tech%20Memo%202.pdf.
3. Federal Highways Administration. *Traffic Bottlenecks Web Page*. https://ops.fhwa.dot.gov/bn/lbr.htm#g9. Accessed July, 2017.
4. Ahanotu, D., R. Margiotta, B. Eisele, M. Hallenbeck, A. Goodchild, and E. McCormack. *NCHRP Research Report 854: Guide for Identifying, Classifying, Evaluating, and Mitigating Truck Freight Bottlenecks*. Transportation Research Board of the National Academies, Washington, D.C., 2017. http://www.trb.org/Main/Blurbs/176115.aspx.
5. Bertini, R. L., and A. M. Myton. Use of Performance Measurement 533 System Data to Diagnose Freeway Bottleneck Locations Empirically in 534 Orange County, California. *Transportation Research Record: Journal of the Transportation Research Board*, 2005. 1925: 48–57.
6. Lund, A., M. Pack, C. Plaisant, and M. Franz. Algorithms for Identifying and Ranking 2 Bottlenecks Using Probe Data. Presented at 96th Annual Meeting of the Transportation Research Board, Washington, D.C., 2017.
7. Chen, C., A. Skabardonis, and P. Varaiya. Systematic Identification of Freeway Bottlenecks. *Transportation Research Record: Journal of the Transportation Research Board*, 2003. 1867: 46–52.
8. Chen-Fu, L. *Using Truck GPS Data for Freight Performance Analysis in the Twin Cities Metro Area*. Report Number MN/RC 2014-14. Minnesota Department of Transportation, Saint Paul, MN, 2014.
9. Wisconsin Department of Transportation. *Wisconsin Sate Freight Plan*. 2017. http://wisconsindot.gov/Pages/projects/sfp/chapters.aspx. Accessed July, 2017.
10. Navjoy Consulting Services, Inc. *CDOT INRIX Validation Project*. Available by request from CDOT, 2016.

The Standing Committee on Statewide Multimodal Transportation Planning (ADA10) peer-reviewed this paper (18-04958).

Article

TRR
JOURNAL OF THE TRANSPORTATION RESEARCH BOARD

Transportation Research Record
2018, Vol. 2672(51) 31–44

DOI: 10.1177/0361198118776154
journals.sagepub.com/home/trr
SAGE

Is Performance Measurement Improving Planning Practice? The Case of Location Affordability in Long-Range Transportation Plans

Ann M. Hartell[1]

Abstract

Transportation planning in the United States is moving to widespread use of performance-based planning methods as new federal requirements for metropolitan planning organizations (MPOs) are implemented. In addition to requirements for measures of safety, infrastructure condition, and travel reliability, many MPOs are adopting additional performance measures including those of complex socioeconomic issues. This study explores current planning practice in using a performance-based approach to tackle such an issue: location affordability, defined as an affordable level of combined cost burdens for housing and transportation. A review of long-range transportation plans at 21 large MPOs provides information on how location affordability is represented in regional transportation plans, how it is defined and measured, and how it is integrated into the planning process. The results are interpreted using a theoretical framework designed to help match planning methods to the conditions of a planning situation. The framework, Christensen's matrix of planning problems, provides insights into the use of performance measures for location affordability and how performance measurement programs can be designed to support more effective planning for this and similarly challenging issues.

Regional transportation planning agencies in the U.S. are called upon to address an array of social issues. Among these is location affordability, or the combined cost burdens of housing and transportation for households. Location affordability is a concept that first garnered national attention in 2006 when the Center for Neighborhood Technology, a non-profit research and advocacy organization, published their influential Housing + Transportation (H + T) Index (*1*). The Index has captured the attention of planning practitioners, stakeholders, and advocacy groups and is an emerging topic in planning efforts.

Resolving unaffordability, however, presents serious challenges, requiring coordination between land development processes, affordable housing programs, transportation policy, and the routine operation of the transportation system. These sectors are subject to wider economic forces and demographic trends over which transportation agencies have little if any control. Still, in many regions, metropolitan planning organizations (MPOs) are responding to stakeholder concerns about affordability and addressing this issue in their regional plans.

At the same time, the processes that produce MPO long-range transportation plans (LRTPs) are transitioning to greater use of performance-based frameworks and methods. Although performance management methods have been used by transportation agencies for some time, federal legislation now requires MPOs to adopt a performance-based approach to planning. These requirements were first established in 2012 with the passage of Moving Ahead for Progress in the 21st Century Act (MAP-21) and were largely continued under the 2015 Fixing America's Surface Transportation (FAST) Act (*2*). Substantively, these requirements focus on safety, infrastructure condition, and travel reliability; however, MPOs are applying performance management methods to other planning issues as they restructure their processes to comply with the new requirements. In a recent

[1]Institute for Multi-Level Governance and Development, Wirtschaftsuniversität Wien/Vienna University of Economics and Business, Vienna, Austria

Corresponding Author:
Address correspondence to Ann M. Hartell: ahartell@nas.edu

survey, two-thirds of MPOs indicated interest in going beyond the federally required measures (*3*).

It is worthwhile to consider some of the implications of this shift. In particular, it is useful to investigate whether performance-based approaches to planning improve MPO capacity to address thorny issues like location affordability. This study explores current practice through a review of LRTPs from 21 large MPOs. The review documents how location affordability is defined and measured, and how it is integrated into performance-based planning processes. This information is then interpreted using Christensen's matrix, a theoretical framework to match planning approaches to the conditions and characteristics of the planning situation (*4*). The results offer insights for the constructive application of performance measures of location affordability, with lessons for similarly complex planning challenges.

Performance Management in Transportation Planning

Performance management is defined as the "process of defining, monitoring, and using objective indicators of the performance of organizations and programs to inform management and decision-making on a regular basis" (*5*, p. 1). Performance management is a collection of mechanisms that are adopted to improve the performance of an organization, including performance measures. More than just published data, a performance measure is a variable, or combination of variables, selected to represent a characteristic or issue of interest and applied in a goal-setting context (*6*). Performance monitoring, the practice of using performance measures in a reporting program, is a prominent, if not defining, feature of performance management (*7*).

In the transportation planning context, performance-based planning and programming (PBPP) is "the application of performance management within the planning and programming processes of transportation agencies to achieve desired performance outcomes" (*8*, p. 1). In addition to supporting performance monitoring and target setting, other purported benefits of PBPP include clarifying goals by converting them to measurable objectives; supporting policy and investment decisions by allowing for systematic comparison of alternative options; and assessing the effectiveness of projects and strategies in furthering progress towards goals (*8*). Efficacy is assessed as the degree to which the implementation of the proposed plan delivers progress towards plan goals.

Any use of performance measures incorporates an element of accountability. This is done through the reporting mechanisms and feedback relationships between an implementing agency and the entities that empower it, either directly or indirectly, to do its work. The strength of accountability can vary; weak accountability is characterized by simply reporting on activity and outputs, while strong accountability will link results to incentives (or sanctions), such as funding levels. In this way, performance management is "governing the government" by regulating the available options for decision-makers and steering their activities (*7*). For an MPO, the accountability environment includes the MPO policy board, local governments, state departments of transportation (DOTs), state legislatures, state and federal regulators, and the public. MAP-21, the FAST Act, and related federal rules formalize the accountability relationship between MPOs and the federal government by requiring performance measures, targets, and reporting on progress towards federal objectives in exchange for the use of federal funds (*9*).

Yet transportation agencies may be reluctant to adopt performance measures—and the accountability this entails—of phenomena over which they have little, limited, or only indirect control (*10*). Second, when there is a lack of clarity about the direction and degree of effect of specific interventions and desired outcomes, an agency may not be able to predict the likely effects of its projects and programs. Or the effects may play out over extended timeframes and be highly subject to confounding effects. Third, numerous benefits flowing from transportation are qualitative or subjective factors which do not lend themselves to tidy quantification. This presents a serious challenge to developing clear and valid measures that can be combined with quantitative traffic or infrastructure measures in a performance measurement system (*10*). Additional technical issues include defining appropriate thresholds and categories of population groups for socioeconomic impacts and benefits (*11*). Although social and economic factors present challenges in a quantitative, performance-focused planning environment, these factors are important considerations for the well-being of individuals, households, communities, and regions. Thus, it is worthwhile to consider how such topics are incorporated into PBPP and LRTPs.

The Case of Location Affordability

As a planning issue, location affordability has several features that make it an appropriate case for examining performance-based approaches to planning. Location affordability raises several technical challenges including identifying a meaningful threshold for affordability. Although the Center for Neighborhood Technology (CNT) proposes a threshold of 45% of household income, this is a normative view that combines the existing housing policy standard of 30% of income for housing affordability and the so-called "attainable goal" of 15% of income on transportation (*12*, p. 3). Further measurement issues include appropriate geographic scale to measure outcomes (region, neighborhood, etc.) and whether affordability is only relevant for certain demographic groups (lower-income, minority, etc.).

Although there may be a regional consensus that location affordability is a priority issue, it may not be clear what MPO actions will deliver desired progress. The forces that affect housing and transportation cost burdens are the result of many actions by governments, developers, households; wider economic forces that affect employment and income; and historic processes that create and sustain social inequities. Designing interventions that will improve affordability requires finding solutions that simultaneously shape the supply and spatial distribution of housing units, housing choices by households, transportation options, and travel behavior, while controlling counterproductive processes such as gentrification and residential displacement. These processes are not under the full control of an MPO. Further, potential solutions may be out of reach because of funding, regulatory, or political limitations. The combination of measurement and technical challenges, and complex causal relationships that muddy attribution of effects, coupled with implementation challenges, introduce considerable uncertainty into any effort to address location affordability into a performance-based approach to the LRTP process.

Methodology and Data

There are several methods for studying plans to document planning practice. For transportation plans, the most common approach is what Baer terms "plan critiques" (*13*). Examples of plan critiques include reports for the FHWA such as *Integration of Context Sensitive Solutions in the Transportation Planning Process* (*14*) and *Metropolitan Area Transportation Planning for Healthy Communities* (*15*). One of the weaknesses of plan critique is the lack of a systematic method and sampling frame; thus, these studies tend to reflect the idiosyncratic interests of the researcher or the project sponsor or both. A few examples of systematic reviews include reviews of accessibility measures in metropolitan plans (*16*), of state DOT sustainability plans (*17*), and of safety in pedestrian master plans (*18*). The overall goal of these studies is to identify patterns, that will, over time improve the quality of plans and increase plan efficacy in addressing real-world issues.

There are a few published studies of how the shift to performance-focused methods is affecting transportation planning practice. For example, Boisjoly and El-Geneidy (*19*) used a qualitative assessment of 32 metropolitan plans to focus on definitional and measurement issues of accessibility in plan objectives and performance measures. Other researchers have focused on the methods and metrics used to assess the equity impacts of transportation plans (*11*, *20*). Yet most published research is focused on the technical dimensions of performance measures (e.g., validity, reliability, and existence of a conceptual linkage to plan goals).

This study provides a systematic review of location affordability in plan narratives, goals, objectives, and performance measures. The review also documents how location affordability measures are used in the planning process and interprets the information using Christensen's matrix as a theoretical framework.

Sample

The review includes plans current as of December 2016 from 21 MPOs, each with populations over 2.5 million in their planning jurisdictions in 2010 (*21*). (The Puerto Rico MPO is excluded from the analysis because a current LRTP was not available online at the time of this analysis.) Altogether, these MPOs are tasked with transportation planning for over 115 million residents, just over 37% of the total U.S. population in 2010.

The plans represent a mix of full length ("major update") LRTPs and shorter ("minor update") plans, depending on where an MPO was in their 4-year plan update cycle. All the plans were developed before the Final Rule on federally required performance measures was in effect. Even so, interest in and preparation for implementing PBPP practices is very much in evidence; several plans describe the upcoming changes in federal requirements and outline their efforts toward compliance.

The plan horizon is 2040 for all plans except Phoenix and Denver which have a 2035 horizon. In some cases, the plans refer to additional analyses or related planning documents, such as plan appendices or stand-alone reports on their performance management program. Some plans are integrated with or draw heavily from a separate policy document, vision plan, or regional comprehensive plan (land use, transportation, housing, and other public services). In such cases, these additional documents were also reviewed.

Plan Review Process

The plans were reviewed in several steps. First, each plan was scanned to gain an overall understanding of the main issues for the region and of the MPO's general approach. Particular attention was given to plan sections on social equity, economic prosperity, livability, and similar, where the issue of location affordability was likely to appear. Using a data collection form, goals, objectives, and strategies for location affordability were noted along with how any location affordability performance measures were applied in the planning process. Next, relevant related documents were added to the review. Finally, a search was made of each plan document for the terms 'afford', 'affordability', 'affordable', 'equity' and 'equitable' to check that all sections where location affordability was discussed had been reviewed.

It is important to note that several plans gave considerable attention to affordable housing, and a few discuss transportation affordability, but do not consider these cost categories as combined cost burdens. This may represent a strategic move to align categories of cost burdens with planning entities (i.e. transportation affordability in a transportation plan, housing affordability in a regional comprehensive plan). Segmenting cost burdens, however, can obscure economic challenges faced by households, perhaps even leading to counterproductive policies and projects such as remote siting of affordable housing where low-cost transportation services are scarce or infeasible (*22*). In the review, information about how plans cover these separate cost burdens is noted; nevertheless, only plans that explicitly linked cost burdens for housing and transportation are considered to be taking up location affordability. Note that the absence of location affordability in the LRTP is not necessarily evidence that a region does not have an affordability issue or is not making efforts to address affordability through some alternative pathway.

This method has some potential drawbacks that may impose limitations on the results. Areas of concern include potential reviewer bias (unreliable reading of the plans), instrumental bias (the review method is unreliable across plans), or the review is incomplete due to missing information from related documents or other sources. These potential sources of bias are somewhat reduced by focusing the review narrowly, compiling information using a data collection form, and making a second round of reviews of those plans that did include a location affordability indicator.

Including only large MPOs may also limit the generalizability of the results. Large MPOs may face particular complexity as a result of a greater number of governmental units in their planning jurisdictions. Large MPOs may also have greater staff capacity and other resources (e.g. data, models) perhaps making them more likely to take up emerging issues or innovative approaches than smaller MPOs (*11*).

Plan Review Results

The 21 regions take various approaches to addressing affordability. Some MPOs limit the issue to a discussion in the plan narrative. In some regions, location affordability is a prominent topic of a related plan, such as a regional comprehensive plan or regional sustainability plan, but is given less attention in the LRTP. For example, although location affordability is not an issue discussed in Baltimore's LRTP, this region developed a regional plan for sustainable development (RPSD) supported by the U.S. Department of Housing and Urban Development's Sustainable Communities Planning Grant program (*23*). The Baltimore RPSD has strong emphasis on strategies to reduce location cost burdens, with \$100 million in funded projects for job accessibility projects and programs. For some MPOs, location affordability is not portrayed as an important issue. This may be because not all regions have a location affordability problem, or it may be a lower priority than other issues. For example, the Detroit plan mentions a need for accessibility to jobs for residents in affordable housing, but the major focus of the plan's solutions is system preservation and transportation funding.

Most plans do not implement a location affordability performance measure in their current LRTPs, despite the issue getting at least passing mention in 13 of the 21 plans. A few plans include discussion of location affordability in connection with their upcoming transition to PBPP, signaling that while location affordability may not be part of a performance-based planning process in the current plan, the MPO anticipates it will be in the future.

Five MPOs include an indicator, target, or performance measure of location affordability in the current LRTP: Chicago, Los Angeles, San Francisco, St. Louis, and Washington, D.C. Two additional plans, Dallas-Fort Worth and Minneapolis-St. Paul, note a location affordability measure under consideration for future plans. Table 1 summarizes the information from the plan review.

Defining and Measuring Location Affordability

Plans that discuss location affordability frequently cite the H + T Index. The regional indicator report for Washington DC refers to the H + T Index as the "national standard for assessing basic community affordability" (*24*, p. 23). Yet adopting the Index as an off-the-shelf performance measure is not without problems. For example, the Chicago region adopted the Index as part of a regional monitoring program in 2010. For the 2017 LRTP update process, the regional indicators are revisited to consider adjustments to improve their validity, their comprehensibility to nontechnical audiences, and to evaluate the quality, reliability, and timeliness of the underlying data. Specific improvements include changes to use "actual measured values" instead of modeled or estimated figures (*25*, p. 2). The MPO identified two issues with the H + T Index: the use of modeled data for

Table 1. Summary of Plan Review

MPO, main city, plan document(s) reviewed, month and year LRTP adopted	LA in goals	LA a strategy or objective	LA in PMs	LA measure	How LA PM is used	Data used for LA PM	Comments
Metropolitan Transportation Commission San Francisco Plan Bay Area July 2013		✓	✓	"Decrease by 10 percentage points (to 56 from 66%) the share of low-income and lower-middle income residents' household income consumed by transportation and housing."	Scenario evaluation, project prioritization; equity analysis.	Regional model outputs; vehicle costs include costs of owning and operating.	In addition to the LA PM, the equity assessment of the plan has a measure of displacement risk: "% of rent-burdened households (50% or more of income) in high-growth areas"; preferred plan scenario does not improve on these 2 indicators; plan narrative calls LA "one of the greatest regional challenges to tackle over the coming years." Housing affordability tracked in regional monitoring program.
East-West Gateway Council of Governments St. Louis Connected 2045 June 2015		✓	✓	"Average proportion of household income spent on housing and transportation costs in the St. Louis region."	Regional monitoring + PM for transportation system	American Community Survey (after the H + T Index)	Use a measure of accessibility for project prioritization that is linked in narrative to LA, but method not reported in plan.
Chicago Metropolitan Agency for Planning Chicago Go To 2040 October 2014	✓		✓	"Percentage of income spent on housing and transportation by moderate-income and low-income residents: 53% by 2015, 45% by 2040."	Regional monitoring	Consumer Expenditure Survey, BLS	Targets referred to as "aspirational goals."

(continued)

Table 1. Continued

MPO, main city, plan document(s) reviewed, month and year LRTP adopted	LA in goals	LA a strategy or objective	LA in PMs	LA measure	How LA PM is used	Data used for LA PM	Comments
Southern California Association of Governments Los Angeles The 2016-2040 Regional Transportation Plan/ Sustainable Communities Strategy April 2016				"Percent of income spent on housing and transportation."	Regional monitoring	BLS; ACS	LA indicator not applied to current plan but will be used for monitoring going forward. Out of pocket transportation costs included in equity assessment.
National Capital Region Transportation Planning Board Washington D.C. 2014 Constrained LRTP October 2014				"By 2020, the housing and transportation costs in Regional Activity Centers will not exceed 45% of area median income."	Regional monitoring	H + T Index	Minor update plan. General note that the MPO will revisit their PM program and may use different data sources or PMs in the future and as MAP-21 requirements are finalized; suggest PMs will be used to select plan scenarios and/ prioritize projects in future plans.
North Central Texas Council of Governments Dallas-Fort Worth Mobility 2040 March 2016				"Relative change in Housing + Transportation Affordability Index."	Potential use in future	H + T Index	LA PM catalogued for potential future implementation in PBPP or regional monitoring.
Metropolitan Council Minneapolis-St. Paul Thrive MSP Transportation Policy Plan January 2015					Potential use in future		Plan catalogs an extensive list of possible measures including an LA PM.
Delaware Valley Regional Planning Commission Philadelphia Connections 2040 Plan for Greater Philadelphia July 2013	✓						Minor update plan; previous plan included measure of household energy and auto costs for scenario analysis; use a housing affordability PM in a regional monitoring program.

(continued)

Table 1. Continued

MPO, main city, plan document(s) reviewed, month and year LRTP adopted	LA in goals	LA a strategy or objective	LA in PMs	LA measure	How LA PM is used	Data used for LA PM	Comments
Puget Sound Regional Council Seattle Transportation 2040 May 2014		✓					Update plan largely focused on revising the financial element and revenue estimates; minor discussion of housing affordability; indicate interest in adopting new PM practices, especially use of more quantitative measures.
Boston Region MPO Boston Charting Progress to 2040 July 2015		✓					Link between objective to minimize cost burdens for households and project evaluation and scenario analysis not evident.

Additional Plans Reviewed

Atlanta Regional Commission, Atlanta	The Atlanta Region's Plan, Transportation Element, February 2016
Baltimore Metropolitan Council, Baltimore	Maximize 2040, November 2015
Denver Regional Council of Governments, Denver	2035 Metro Vision Regional Transportation Plan, February 2011
Houston-Galveston Area Council, Houston	Bridging Communities 2040 RTP, March 2016
Maricopa Association of Governments, Phoenix	2035 Regional Transportation Plan, January 2014
Miami-Dade MPO, Miami	Eyes on the Future, October 2014
New York Metropolitan Transportation Council, New York	Plan 2040, April 2016
North Jersey Transportation Planning Authority, Newark	Plan 2040, September 2013
San Diego Association of Governments, San Diego	San Diego Forward, October 2015
Southeast Michigan Council of Governments, Detroit	Creating Success with Our Transportation Assets, June 2013
Southwestern Pennsylvania Commission, Pittsburgh	Mapping the Future: A Southern PA Plan, June 2015

Note: LA = location affordability; PM = performance measure; BLS = Bureau of Labor Statistics; ACS = American Community Survey

transportation costs and of multi-year estimated data from the American Community Survey for most other variables (*25*). These characteristics make the Index less useful for programs seeking to track year over year change in observable outcomes. Therefore, the MPO elected to use annual Consumer Expenditure Survey (CES) data to calculate a location affordability indicator. However, this change will introduce a new disadvantage for use in PBPP because CES data are reported at the metro level while transportation and housing interventions are most often site- or corridor-specific. As a result, an effective, but small-scale intervention could be undervalued by the region-wide indicator.

The scale of measurement can also mask considerable disparities among socioeconomic groups. Several of the plans tackle this issue in equity assessments. For example, the St. Louis MPO models subregional variation of cost burdens, finding a different picture than the relative affordability for the region as a whole. The analysis focuses on differences in transportation cost burdens, which can be projected using the MPO's models (*26*). For environmental justice (EJ) populations, the location affordability indicator is 52% of income at current gas prices. Modest cost burdens for the region as a whole are masking high vulnerability for EJ populations, largely as a result of a significant gap in median income ($35K vs. $62K). The disparity in income is coupled with relatively low transit use levels, which might offer a lower cost option for lower-income households (*26*, *27*). Although the disparities do not rise to a formal standard of a "disproportionate impact", they demonstrate how disaggregated data can reveal considerable differences within a region. The Washington DC MPO reports a similar pattern when disaggregated data are compared with regional values (*24*, p. 2).

The MPO for the Los Angeles region uses two cost burden indicators in their plan. The first is the percent of income spent on both housing and transportation, calculated using BLS and ACS data. This indicator is part of a regional monitoring program and an overall plan coordination effort, but is not directly used to measure the benefits of the LRTP because of the difficulty of projecting housing costs (*28*). The second indicator is annual household costs for transportation only, using the same method as for the transportation cost component of the H + T Index. Here, the preference for modeled data is in order to project potential benefits of the plan. The San Diego region MPO also uses modeled transportation cost burdens in its equity assessment but does not link them with housing costs.

Apart from the frequent mention of the H + T Index, no strong pattern in the approach to measuring location affordability emerges. Some MPOs focus on high-level, regional indicators; others assess subregional differences in affordability. MPOs also differ in their use of modeled and observed data. The Chicago experience provides an example of internal critique and review to maintain a performance measurement program, and highlights a potential drawback to adopting an indicator that is readily available but not well-aligned with agency needs.

Accountability Environment

Looking across the plans there is a general characteristic of measuring location affordability in a weak accountability framework. That is, while a location affordability indicator is implemented, if an agency fails to adequately perform, there are no ramifications, such as a reduction in funding. For example, in the Washington DC, and Chicago regions, location affordability is included in a regional monitoring program and has some link to LRTP goals and objectives but this does not yet rise to a performance-based approach to planning. Instead, these examples are better described as a form of performance communication.

The St. Louis region also has a regional monitoring program, but the LRTP builds on this by including location affordability in its equity assessment of its plan, thus accepting a somewhat higher degree of accountability for affordability effects of the plan for EJ populations. The plans for San Diego and Los Angeles are similar; cost burdens are included in their equity assessments, but these are limited to transportation cost burdens. The San Francisco region plan includes a location affordability measure in its equity assessment as well as in an assessment of the overall impacts of the plan. These assessments, however, do not find that the plan is expected to reduce housing and transportation cost burdens. In fact, projections of the adopted plan scenario move in the wrong direction, driven by projected increases in gas prices, a housing market recovery expected to increase housing prices, and population growth. The plan's shortfall is used as an opportunity for performance communication. In a frank discussion, the difficulty of the region's affordability dilemma is described along with the limits of the ability of the MPO to resolve the issue and some explanations as to why that is the case, observing that greater affordability remains "vexingly out of reach" (*29*, p. 61).

Location Affordability in the Planning Process

In PBPP, performance measures can be applied in several ways: as criteria for selecting or prioritizing projects, for comparing plan scenarios, or for evaluating the expected benefits of an adopted plan. These assessments can be carried out for the whole region, for subareas, or for selected demographic groups. The plans reviewed reflect these diverse applications.

In the Chicago region, affordability is incorporated into a project-level assessment with an evaluation of how specific projects are expected to change the number of affordable housing units in transit-oriented developments and overall increases in density (*30*, *31*). The Washington DC MPO also uses a project-level assessment of major projects with qualitative scoring of location affordability, charting which transportation goals from the policy plan are supported for each project (*32*, p. 6). This MPO also uses a quantitative location affordability indicator in a regional monitoring program (*33*).

In the St. Louis plan, the location affordability measure is applied in the plan analysis at the scale of the entire plan but geographically disaggregated to smaller geographies (traffic analysis zones) and for demographic subgroups in an equity analysis (*26*). However, there is no evidence that this information was used to test alternative plan scenarios or the potential benefits of proposed projects. A second, project-level assessment, described in the Transportation Improvement Program (TIP), is a qualitative scoring of how well a project improves access to opportunity for low-income and disabled individuals (*34*). The scoring process used is not described in detail, but the discussion in the TIP suggests an ordinal scale. The change in emphasis from the plan level to the project level indicates a disconnect that may weaken the linkage between plan goals and investment priorities despite there being a performance measure for affordability.

As noted previously, the Los Angeles region includes a location affordability in a regional monitoring program; however, this indicator is not used in the plan assessment because of the difficulty of projecting housing costs (*28*, p. 23). In the equity assessment of the LRTP, only a transportation cost burden measure is applied.

The San Francisco region's plan describes a somewhat higher degree of integration of a location affordability indicator into the planning process, applying it at multiple points in the plan development process. Performance targets are used to evaluate the potential pairings of land use scenarios and transportation system scenarios in an iterative fashion. Repeated assessments of the equity effects of the plan scenarios also were made during the plan development process (*35*). Then the preferred plan scenario was assessed with respect to goals and targets for the plan. The target for location affordability is a 10% reduction in housing and transportation cost burdens for low- and lower-middle income households, which would bring the region in line with the national average (*29*). The equity assessment of the plan compared changes in these cost burdens for vulnerable populations compared with the rest of the region. An indicator of displacement risk, defined as the share of renter households with high housing cost burdens in areas where high growth is anticipated, is also reported.

The San Francisco plan reveals a difficult dilemma for the region and for MPO capacity to address location affordability. As for the other two California plans, the San Francisco plan was produced under California SB 375 which mandates reductions in greenhouse gas (GHG) emissions and for planning for adequate housing to accommodate future growth (*29*). The GHG reduction targets require planning that will trigger shifts to lower-carbon travel modes and shorter commutes. The requirement to fully accommodate future housing demand requires planning for increasing density and infill development. The plan responds to these requirements with an overall strategy of concentrating growth, increasing the housing supply in existing communities, and upgrading and expanding transit service. These planned interventions are not, however, projected to be able to offset high cost burdens for lower-income households. In fact, these strategies may exacerbate the problem: improved transit service may make housing units close to transit even more costly while geographically constraining new housing construction may increase housing market pressures. The results of the plan assessment seem to bear this out. While the plan offers a plausible pathway to meeting state GHG and housing targets, it is unable to simultaneously improve on affordability. Here, although location affordability is integrated into a performance-focused approach, other goals, outside factors, and mandated targets had greater influence over the plan.

Summary

The review of plan documents for 21 large MPOs finds 6 plans (28%) include location affordability in the goals, objectives, or strategies of their LRTPs. Four regions (19%) include an indicator of location affordability in their regional monitoring programs (Chicago, Los Angeles, St. Louis, Washington D.C.). Two regions include a location affordability indicator in a list of indicators for potential adoption in the future (Dallas-Fort Worth and Minneapolis-St. Paul). In 4 regions—San Francisco, San Diego, Los Angeles, and St. Louis—a measure of transportation costs is used in the equity assessment. A few plans use a qualitative approach to assessing projects. Only one plan, for the San Francisco region, is found to use a quantitative location affordability measure as part of a performance-based planning effort to develop an LRTP, not only as a test of likely results from a selected plan scenario or of specific projects.

In discussing, defining, and measuring location affordability, the plans draw heavily from the H + T Index. Plan narratives frequently mention the Index, and the 45% affordability threshold proposed by CNT is adopted by some MPOs as a target or benchmark. The H + T Index has clearly had an impact on the work of MPO planners, although its use as a performance measure in PBPP may have limited potential.

Overall, indicators for location affordability are applied in a weak accountability environment. Regional monitoring programs focus on communicating information about affordability and do not link progress (or lack thereof) to incentives (or penalties). Nevertheless, there is a degree of accountability in including affordability indicators in an equity assessment as part of meeting the requirements of Title VI of the Civil Rights Act to demonstrate that an MPO plan will not disproportionately burden or deny benefits to low-income or minority populations (*36*).

Christensen's Matrix: A Theoretical Framework for Interpretation

A key component in effective planning is the design of a planning process that is appropriate for the overall planning situation. Christensen (*37*) developed a matrix to classify planning situations according to conditions of uncertainty. Christensen's matrix plots uncertainty along two axes: goals and solutions (see Figure 1). Goals are "uncertain" in the sense that there can be less than full consensus for a goal, or there may be multiple goals that have strong support but are in conflict. Solutions are those interventions that advance the goals. In transportation planning, solutions can be policies, programs, or projects. Solutions are "certain" to the degree they are replicable and known to deliver dependable results. The central idea of the matrix is that by clarifying the nature of uncertainty, planners can improve their effectiveness by aligning planning processes with the task at hand. As a planning tool, quantitative measures can be used in any planning circumstance, but their purpose, design, and application must vary to support effective planning.

Christensen argues that failing to recognize uncertainty and behaving as though issues and solutions are clear will lead planners to produce ineffective plans or even implement projects with harmful unintended effects. "Delusions of certainty" in the solutions dimension encourage adoption of policies that incorrectly assume an understanding of the connection between means and ends. Here, expert knowledge may be overestimated, and disciplinary biases may presuppose a particular solution set. When uncertain solutions are applied, the results are often surprising, uneven, or disappointing. On the goal dimension, failing to diagnose uncertainty may lead to an artificial narrowing of issues so they match preferred solutions of an agency or particular discipline.

When there is consensus on goals and solutions are known (the upper left quadrant), the issue can be tackled with routinized processes of government, through bureaucratic operations that deliver replicable, dependable results. In these circumstances, performance measures work well as measures of organization efficacy in delivering projects and programs that will lead to desired outcomes.

In the upper right quadrant, there is disagreement or conflict among stakeholders on the goals. For each position in the debate, advocates can point to a known technology and how that technology would support their preferred goal. Until the lack of agreement on what goals are to be addressed is resolved, selecting a solution is expected to lead to conflicts among stakeholder groups. In this quadrant, the appropriate approach is a bargaining process through which the goals themselves are clarified; the planning agency can support a successful political bargaining process that accommodates various preferences or a tradeoff analysis that balances constraints and competing interests (*37*). Because outcomes of bargaining process are highly dependent on the particular actors involved and, thus, unpredictable, it is inappropriate to expect consistent, replicable outcomes. However, agencies can be held accountable for their contributions to the quality and inclusiveness of the debate. Thus, appropriate performance measures relate to the enforcement of procedural rules that foster productive debate and broaden access to the decision-making process. Also appropriate are measures of effectiveness in enforcing any agreed-upon mitigation (*37*).

The lower left quadrant represents situations where there is consensus on a goal, but the means to achieve it are unknown or unproven. In these cases, resolving the uncertainty requires searching for more information through pilot programs, phased implementation, experimentation, and innovation. Without a proven solution, it is inappropriate to hold agencies accountable for results (*37*). Nevertheless, agencies can be held accountable for collecting and assessing information about outcomes from any candidate solutions that are implemented.

The lower right quadrant represents those situations where there is uncertainty about solutions and lack of agreement on goals. This is a zone of chaos and of "wicked problems" (*38*). In this zone, goals are vague or conflicted, facts are debated, competing positions offer no proposals for solutions, and there may be inadequate levels of trust to support a bargaining process. Planning tends to become reactive and undirected (*4*). Resolving the situation requires shifting it to another quadrant. This can be done by restructuring or redefining the problem to one that participants agree can be worked on,

Goals → Solutions ↓	Consensus	No Consensus
Known	*Bureaucracy & Stable Rules* Replicable solution(s) Dependable results Perf. measure applications: • Measure efficacy in program delivery • Measure value of program benefits	*Pluralist Politics & Agreed Rules of the Game* Proven technologies available for various potential goals Bargaining and trade-offs Perf. measure applications: • Measure efficacy in following procedural rules • Measure enforcement of terms of bargaining agreements
Unknown	*Change & Rules of Experimentation* Shared commitment to a problem/goal No proven solution(s) Pilots, experimentation Perf. measure applications: • Measure effects of programs including differences by context • Measure agency support of innovation and experimentation	*Wicked Problems & Problem Finding* Conflicted or unclear goals Unproven solutions Zone of chaos Perf. measure applications: • Measures of leadership • Measures of solution identification • Measures of effectiveness in restructuring and formation of multiple goal positions

Adapted from Christensen (*4, 37*).

Figure 1. Christensen's Matrix.

which moves the situation to the lower left quadrant. Alternatively, the debate can be restructured to focus on problems that can be solved with existing solutions, which moves the situation to the upper right quadrant. The application of performance measures or even evaluation frameworks in this chaotic quadrant is limited (*39*).

Performance Measures and the Reduction of Uncertainty

With respect to location affordability, there is no evidence that MPOs are in the upper left quadrant of Christensen's matrix where unaffordability can be resolved by administering a routine program that delivers reliable results that improve location affordability. Thus, the application of performance measures in their traditional sense, as measures of organizational effectiveness in delivering replicable results, is inappropriate.

For some MPOs, the situation may be in the upper right quadrant, where the location affordability goal is in conflict with other goals. The California mandates for adequate housing and GHG emissions reductions may locate the planning situation for location affordability to the upper right quadrant, fixing these goals and leaving little space for resolving goal conflict through pluralist politics and bargaining. Alternately, a location affordability goal may be adopted as a proxy for an overall equity goal. If so, the affordability goal may in fact be a "charter goal." Charter goals serve the function of creating a common ground and a community, but are "more like a receding horizon" than targets for attainment (*40*, p. 536). However, communicating a regional monitoring program focused on charter goals as though it is a robust performance measurement framework can create unrealistic expectations and fuel negative perceptions of agency efficacy when problems persist.

An additional area of uncertainty is in what solutions are effective in reducing unaffordability. This is likely for location affordability; even carefully crafted plans with an explicit affordability goal are unable to make much headway on the issue. In the plans reviewed, there is a general pattern of recommending transit, transit-oriented development, and preserving affordable housing units near transit as ways to improve location affordability. These are not particularly new or innovative ideas, but are apparently inadequate for resolving regional affordability challenges and plans acknowledge their potential for unintentional consequences including gentrification and displacement. Overall, the plans reveal considerable uncertainty in solutions for unaffordability even if there is consensus on location affordability goal.

If location affordability is an issue that can be sited in the lower left quadrant of Christensen's matrix, where goals are agreed but solutions unproven, performance measures should focus on holding the MPO accountable for information about outcomes, but not for the outcomes themselves. Instead, increasing focus on evaluation and information gathering should test assumptions about the effectiveness of conventional solutions and reduce the tendency toward premature programming. This involves mobilizing performance measures as information-gathering, evaluative mechanisms to learn more about results from various implementation activities that treat projects and programs as experiments or pilots. Yet measuring the effectiveness of a location affordability intervention such as a transit investment or transit-oriented development project requires not regional-level monitoring, but rather fine-grained data, localized analyses, pre- and post-project evaluations, and investigation of disappointing or surprising outcomes. In line with this, Bills and Walker (*20*) suggest harnessing the ability of activity based models (in use or under development by many MPOs) to make detailed demographic analyses and compare plan effects for particular types of households with effects for the region overall. The indicators of gentrification and displacement risk, described as under development in the Los Angeles and San Francisco regions, may also provide important information about unintended effects on affordability from transportation investments.

In recent years, there has been considerable research into the technical aspects of performance measures: data sources, analysis methods, and performance reporting techniques. These will continue to be important aspects of developing a performance-focused approach to planning. However, assessments of the institutional dimensions of planning practice are also needed, and returning to theoretical considerations provides deeper insights into planning practice. In this regard, potential extensions of the current study include expanding the sample of MPOs or assessing changes in practice over time by tracing changes in sequential plans. Examining how MPOs are addressing issues other than location affordability in PBPP could produce further information on emerging patterns. Augmenting a plan review with information from other sources, especially interviews with MPO staff and stakeholders, and working group documents and discussions could further enrich the analysis.

The way that MPOs interpret and use the information from performance measurement programs will largely determine whether PBPP actually strengthens plans and planning practices or becomes a gesture with little connection to decision-making or real-world outcomes. The degree to which performance measurement programs contribute to reducing uncertainty is an important factor for success, especially in connection with difficult issues where goals may be conflicted and solutions unproven. Thus, a performance measurement program should be designed with the conditions of uncertainty in mind to keep the focus on how to better solve problems, not simply on how to better measure them.

Acknowledgments

The author thanks Dr. Ralph R. Hall, Virginia Polytechnic and State University, for his valuable comments in the development of this study. Thanks also to Dr. Gunther Maier and Gillian Foster at the Wirtschaftsuniversität Wien for constructive critiques of a previous version. Additional improvements were suggested by anonymous reviewers from TRB's Standing Committee on Metropolitan Policy, Planning, and Processes (ADA20).

References

1. Center for Neighborhood Technology. *Housing + Transportation Index*. Center for Neighborhood Technology. http://htaindex.cnt.org. Accessed July 18, 2016.
2. Federal Highway Administration. *FHWA Publishes Performance Management Final Rules and NPRMs Required by MAP-21*. Federal Highway Administration. https://www.fhwa.dot.gov/tpm/rule.cfm. Accessed December 2, 2016.
3. Transportation for America. *Transportation Performance Measures, 2017 Survey*. Washington D.C., 2017.
4. Christensen, K. S. Coping with Uncertainty in Planning. *Journal of the American Planning Association*, Vol. 51, 1985, pp. 63–73.
5. Poister, T. H., M. P. Aristigueta, and J. L. Hall. *Managing and Measuring Performance in Public and Nonprofit Organizations. An Integrated Approach*. Wiley, San Francisco, CA, 2015.
6. Gudmundsson, H., R. P. Hall, G. Marsden, and J. Zietsman. *Sustainable Transportation. Indicators, Frameworks, and Performance Measurement*. Samfundslitteratur, Frederiksberg, Denmark, 2015.
7. Triantafillou, P. The Political Implications of Performance Management and Evidence-Based Policymaking. *The*

American Review of Public Administration, Vol. 20, No. 10, 2013, pp. 1–15.
8. Grant, M., J. D'Ignazio, A. Bond, and A. McKeeman. *Performance Based Planning and Programming Guidebook*. Report No. FHWA-HEP-13-041. Federal Highway Administration, U.S. Department of Transportation, Washington, D.C., 2013, p. 133.
9. United States General Accounting Office. *GAO-15-217 - SURFACE TRANSPORTATION - DOT is Progressing toward a Performance-Based Approach, but States and Grantees Report Potential Implementation Challenges*. United States General Accounting Office, Washington, D.C., 2015.
10. Meyer, M. Measuring That Which Cannot Be Measured–At Least According to Conventional Wisdom. In *Performance Measures to Improve Transportation Systems and Operations: Conference Proceedings 26*, Washington, D.C., National Academies Press, 2001. Available at: http://onlinepubs.trb.org/onlinepubs/conf/reports/cp_26.pdf.
11. Karner, A. Planning for Transportation Equity in Small Regions: Towards Meaningful Performance Assessment. *Transport Policy*, Vol. 52, 2016, pp. 46–54.
12. Center for Neighborhood Technology. *H + T Index Methods*. Center for Neighborhood Technology. 2015. https://htaindex.cnt.org/about/HT_Index_Methods_2013.pdf
13. Baer, W. C. General Plan Evaluation Criteria: An Approach to Making Better Plans. *Journal of the American Planning Association*, Vol. 63, No. 3, 1997, pp. 329–344.
14. Center for Transportation and the Environment. *Integration of Context Sensitive Solutions in the Transportation Planning Process*. Center for Transportation and the Environment, Washington, D.C., 2007.
15. Lyons, W., H. Peckett, L. Morse, M. Khurana, and L. Nash. *Metropolitan Area Transportation Planning for Healthy Communities*. Federal Highway Administration, Washington, D.C., 2012.
16. Proffitt, D., K. Bartholomew, R. Ewing, and H. Miller. Accessibility Planning in American Metropolitan Areas: Are We There Yet? Urban Studies, 2017.
17. Mansfield, T., and A. Hartell. Institutionalizing Sustainability at the Level of State Departments of Transportation. *Transportation Research Record: Journal of the Transportation Research Board*, 2012. 2271: 9–18.
18. Jones, D. K., K. R. Evenson, D. A. Rodriguez, and S. A. Aytur. Addressing Pedestrian Safety: A Content Analysis of Pedestrian Master Plans in North Carolina. *Traffic Injury Prevention*, Vol. 11, No. 2, 2010, pp. 57–65.
19. Boisjoly, G., and A. M. El-Geneidy. How to Get There? A Critical Assessment of Accessibility Objectives and Indicators in Metropolitan Transportation Plans. *Transport Policy*, Vol. 55, 2017, pp. 38–50.
20. Bills, T. S., and J. L. Walker. Looking Beyond the Mean for Equity Analysis: Examining Distributional Impacts of Transportation Improvements. *Transport Policy*, Vol. 54, No. 2016, 2017, pp. 61–69.
21. U.S. Department of Transportation. *Metropolitan Planning Organization (MPO) Database*. U.S. Department of Transportation. https://www.planning.dot.gov/mpo.asp. Accessed July 21, 2016.
22. Ewing, R., and S. Hamidi. How Affordable is HUD Housing? *Housing Policy Debate*, Vol. 26, No. 3, 2016, pp. 437–455.
23. US Department of Housing and Urban Development. *Sustainable Communities Regional Planning Grants*. US Department of Housing and Urban Development. https://portal.hud.gov/hudportal/HUD?src=/program_offices/economic_development/sustainable_communities_regional_planning_grants. Accessed March 14, 2017.
24. Hand, R. *Region Forward Coalition Baseline Progress Report*. Metropolitan Washington Council of Governments, Washington, D.C., undated.
25. Chicago Metropolitan Agency for Planning. *Go To 2040 Update Appendix. Indicator Methodology*. Chicago Metropolitan Agency for Planning, Chicago, IL, 2015.
26. East-West Gateway Council of Governments. *State of the System 2045. Technical Supplement to the Long-Range Transportation Plan for the St. Louis Region*. East-West Gateway Council of Governments, St. Louis, MO, 2015.
27. East-West Gateway Council of Governments. *Connected 2045*. East-West Gateway Council of Governments, St. Louis, MO, 2015.
28. Southern California Association of Governments. *Plan Performance. Performance Measures*. Southern California Association of Governments, Los Angeles, CA, 2016.
29. Metropolitan Transportation Commission. *Plan Bay Area. Final Performance Assessment Report*. Metropolitan Transportation Commission, San Francisco, CA, 2013.
30. Dean, B. *Major Transportation Capital Project Evaluation Measures. (Memorandum)*. Chicago Metropolitan Agency for Planning, Chicago, IL, 2009. http://www.cmap.illinois.gov/documents/10180/31056/BoardMemo–CapitalProjects06-10-09.pdf/2cc054fa-23a5-4420-ae6a-00a5a94658eb. Accessed March 4, 2017.
31. Chicago Metropolitan Agency for Planning. *DRAFT Preferred Regional Scenario. An Interim Product of the Go To 2040 Plan*. Chicago Metropolitan Agency for Planning, Chicago, IL, 2009.
32. Austin, A. *2016 CLRP Amendment*. Metropolitan Washington Council of Governments, Washington, D.C., 2016.
33. Swanson, J., and S. Ritacco. *Briefing on the Performance Analysis of the Draft 2016 CLRP Amendment*. Metropolitan Washington Council of Governments, Washington, D.C., 2016. https://www.mwcog.org/assets/1/28/10072016_-_Item_4_-_2016_CLRP_Performance_Analysis.pdf. Accessed March 7, 2017
34. East-West Gateway Council of Governments. *Transportation Improvement Program Fiscal Years 2017 Through 2020*. East-West Gateway Council of Governments, St. Louis, MO, 2016.
35. Metropolitan Transportation Commission. *Plan Bay Area*. Metropolitan Transportation Commission, San Francisco, CA, 2013.
36. FHWA Office of Planning Environment & Realty. *An Overview of Transportation and Environmental Justice*. https://www.fhwa.dot.gov/environment/environmental_justice/overview. Accessed March 4, 2017.
37. Christensen, K. S. *Cities and Complexity: Making Intergovernmental Decisions*. Sage, Thousand Oaks, 1999.

38. Rittel, H. W. J., and M. M. Webber. Dilemmas in a General Theory of Planning. *Policy Sciences*, Vol. 4, 1973, pp. 155–169.
39. Head, B. W., and J. Alford. Wicked Problems: Implications for Public Policy and Management. *Administration & Society*, Vol. 47, No. 6, 2015, pp. 711–739.
40. Landau, M. On the Concept of a Self-Correcting Organization. *Public Administration Review*, 1973, Vol. 33, No. 6, pp. 533–542.

The Standing Committee on Metropolitan Policy, Planning, and Processes (ADA20) peer-reviewed this paper (18-01035).

The author is currently affiliated with the National Cooperative Highway Research Program (NCHRP). The views in this paper are those of the author and do not reflect those of the NCHRP or the Transportation Research Board.

Article

TRR
JOURNAL OF THE TRANSPORTATION RESEARCH BOARD

Transportation Research Record
2018, Vol. 2672(51) 45–55

DOI: 10.1177/0361198118796397
journals.sagepub.com/home/trr
SAGE

An Integrated, Validated, and Applied Activity-Based Dynamic Traffic Assignment Model for the Baltimore-Washington Region

Lei Zhang[1], Di Yang[1], Sepehr Ghader[1], Carlos Carrion[1], Chenfeng Xiong[1], Thomas F. Rossi[2], Martin Milkovits[2], Subrat Mahapatra[3], and Charles Barber[4]

Abstract

The paper discusses the integration process and initial applications of a new model for the Baltimore-Washington region that integrates an activity-based travel demand model (ABM) with a dynamic traffic assignment (DTA) model. Specifically, the integrated model includes InSITE, an ABM developed for the Baltimore Metropolitan Council, and DTALite, a mesoscopic DTA model. The integrated model simulates the complete daily activity choices of individuals residing in the model region, including long-term choices, such as workplace location; daily activity patterns, including joint household activities and school escorting; activity location choices; time-of-day choices; mode choices; and route choices. The paper describes the model development and integration approach, including modeling challenges, such as the need to maintain consistency between the ABM and DTA models in terms of temporal and spatial resolution, and practical implementation issues, such as managing model run time and ensuring sufficient convergence of the model. The integrated model results have been validated against observed daily traffic volumes and vehicle-miles traveled (VMT) for various functional classes. A land-use change scenario that analyzes the redevelopment of the Port Covington area in Baltimore is applied and compared with the baseline scenario. The validation and application results suggest that the integrated model outperforms a static assignment-based ABM and could capture behavioral changes at much finer time resolutions.

Home to 5.8 million people, the state of Maryland is in the center of the eastern seaboard of the United States. Within the state's boundaries is the heavily-congested Baltimore-Washington metropolitan region. Over time, the dense land use, economic activity, over-saturated conditions, and limited capabilities for system expansion have presented unique challenges for mobility, reliability, environmental, and economic vitality. The Maryland State Highway Administration (SHA) and the Baltimore Metropolitan Council (BMC) recognize the value of an integrated, multi-modal, safe, reliable, and efficient transportation system to support and sustain economic growth in this region. Advanced modeling methods and products—such as an integrated, advanced travel demand model and a fine-grained, time-sensitive network that can operate at the metropolitan level—are required by SHA and BMC to meet agency objectives and address various key challenges. Working with SHA and BMC, the University of Maryland proposed to integrate InSITE, an activity-based travel demand model (ABM) developed for BMC by Cambridge Systematics, Inc., with DTALite, a mesoscopic dynamic traffic assignment (DTA) model. With disaggregated data and behavior models, the interactions between supply, demand, policy, and external factors will be better understood.

With advances in both statistical modeling methods and computational technologies, a new generation of travel demand and network supply models has been proposed to support informed decision making and address new policy and investment modeling challenges. In travel demand modeling, ABMs advance the traditional trip-based, four-step models in the following aspects: (a) incorporating the continuous temporal dimension; (b) emphasis on individual-level travel choices; (c) treatment

[1]Department of Civil and Environmental Engineering, University of Maryland, College Park, MD
[2]Cambridge Systematics, Inc., Cambridge, MA
[3]Maryland State Highway Administration, Baltimore, MD
[4]Baltimore Metropolitan Council, Baltimore, MD

Corresponding Author:
Address correspondence to Lei Zhang: lei@umd.edu

of tour-based travel patterns; and (d) capturing intra-household interactions (*1, 2*). On the supply side, DTA models improve the traditional static assignment procedures by: (a) considering time-dependent interactions of the travel demand and network supply; (b) micro-simulation of agents; (c) capturing congestion build-up and dissipation; and (d) ability to incorporate demand management strategies and intelligent transportation system technologies (*2, 3*). The development of integrated ABM and DTA model systems has been of great interest to transportation modelers, since both models employ micro-simulation approaches and could operate at finer time resolutions.

Pendyala et al. (*4*) categorized the ABM and DTA integration approaches into the sequential integration and several levels of dynamic integrations, based on the data exchange frequency between the two models. In a sequential integration paradigm, the ABM and DTA models are loosely coupled, and only communicate and exchange data at the end of a full iteration (i.e., the entire 24-hour period of a day). This sequential information exchange procedure fails to capture the impacts of network disruptions or real-time information systems on travel behavior and demand. In recognition of this limitation, a tighter integration approach is proposed. In a dynamic integration paradigm, data exchange between the ABM and DTA models occurs at a much finer time resolution (e.g., one minute), which represents network dynamics in a more realistic way. Existing dynamic integrated systems include MATSim (*5*), SimTRAVEL (*6*), and POLARIS (*7*). While dynamic integrated modeling systems provide enhanced modeling capabilities, the majority of the integration efforts in the literature falls in the sequential integration paradigm (*8–10*).

Within the scope of this paper, it is not feasible to provide a comprehensive review of the ABM and DTA integration literature. Consequently, this section focuses on reviewing implementation-ready ABM and DTA integrated model systems at the metropolitan level. Particularly, several SHRP 2 Integrated Dynamic Travel Model (C10) funded projects are reviewed in detail. In project C10A, an ABM, DaySim, a DTA model, TRANSIMS, and an emission model, MOVES (the Motor Vehicle Emission Simulator) are integrated for the Jacksonville metropolitan area in Florida (*11*). In project C10B, DaySim and MOVES are integrated with another DTA model, DynusT, for the Sacramento metropolitan area in California (*12*). Other on-going C10 funded projects include the Atlanta Regional Commission's (ARC) project that integrates CT-RAMP1 with DynusT; an Ohio Department of Transportation (ODOT) project that integrates CT-RAMP2 with DynusT; San Francisco County Transportation Authority's project that integrates CHAMP with FastTrips; and Baltimore Metropolitan Council's (BMC) project that integrates TourCast with DTALite. Even though different ABM and DTA models are integrated, the core integration methodologies of these projects are similar and belong to the sequential integration paradigm, except that a tighter integration was proposed for ARC and ODOT. The main challenges stemming from these projects include: (a) excessive runtime: each full model iteration could take days to run; (b) convergence challenge: there is no consensus on unified convergence criteria; and (c) an efficient data exchange procedure is needed to improve the model communication efficiency in order to save model runtime. This paper intends to tackle these challenges while integrating InSITE with DTALite. Technical details are provided in the following sections.

The primary objective of this project is to develop a working model of the Baltimore-Washington region that is disaggregated for auto travel on both the demand and supply side. This model is intended to be application-ready, meaning that it has been validated and includes all functionality required for typical planning analyses. Therefore, the integration pursued between the ABM and DTA model falls in the sequential integration paradigm. In this paradigm, the travel demand model (InSITE) and the DTA model (DTALite) are run in iterations, with one single "big loop" consisting of an iteration of the travel demand model followed by a run of the DTA model. The contributions of this study, in terms of aforementioned integration challenges, are summarized as follows. First, the integrated model runs in an acceptable time frame. With about five million travelers in the system, the average runtime of the integrated InSITE–DTALite model is 24 hours per big loop iteration with 40 iterations of DTA runs. Additionally, a stratified sampling approach is developed to further improve the runtime issue. Second, new convergence criteria based on the number of infeasible agents are proposed to supplement traditional schedule gap-based convergence measures. Third, the ABM and DTA models exchange information through one single trip roster file, which is more efficient than the multiple file exchange (e.g., person file, tour file, and trip file) process in past projects. Finally, this model integration exercise in the Baltimore-Washington region provides implementation experience that is valuable to both researchers and practitioners.

The remainder of the paper is organized as follows. The second section presents an overview of the ABM and DTA models and introduces key modules in these two models. In the third section the integration methodology and technical innovations are carried out in detail. Validation process and results are discussed in the fourth section. Application results of the integrated model to the redevelopment of a brownfield site in Baltimore are analyzed and compared with the baseline

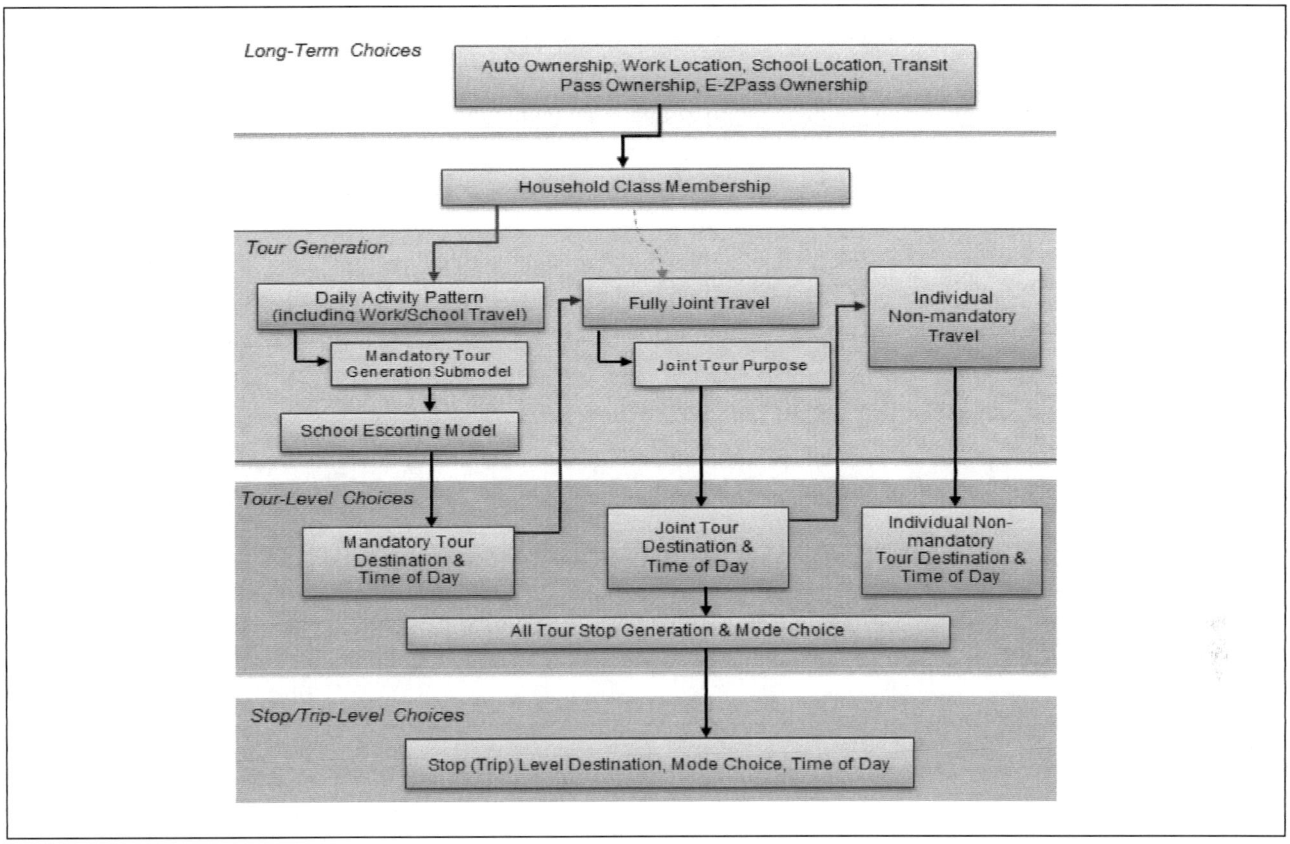

Figure 1. InSITE model structure.

scenario. Finally, conclusions and directions for future research are compiled at the end of the paper.

Overview of Model Components

BMC InSITE ABM

InSITE is an ABM system composed of interconnected, discrete choice models representing choices at distinct dimensions (e.g., travel mode, destination) that focus on decisions related to daily activity and mobility for a typical weekday. InSITE adopts the day activity-schedule approach, where a day activity schedule is defined through the concepts of activity pattern and activity schedule. The activity pattern defines the participation in activities as primary and secondary. Primary activities are the anchors (e.g., home-to-work trip and work-to-home trip represent a tour, with work as the primary activity) of a tour, and secondary activities are intermediate stops within a particular tour (e.g., stopping for shopping during the work-to-home half of the tour). The activity schedule adds detailed information to the activity pattern about tours, such as timing, travel mode, destination of primary activity, and the stops for secondary activities within the tours.

The model covers an area including the entire BMC region, plus the District of Columbia and the Maryland portion of the region covered by the Metropolitan Washington Council of Governments. InSITE models travel for a typical weekday. The choices made by households and individual travelers are simulated using probabilities from a series of logit models. As shown in Figure 1, the model begins by simulating long-term choices that are made before the travel day, including auto ownership, workplace location for workers, school location for students, transit pass ownership, and E-ZPass toll transponder ownership.

The next set of models estimates a daily activity pattern for each person. Whether the person has work tours (with or without stops), school or university tours, non-mandatory activity tours only, or does not travel within the region, is simulated. If the simulated pattern has mandatory (work or school) tours, they are generated. For students making school tours, InSITE simulates whether they are escorted by a household member and if so, by whom. At this point, the destinations and times of day for the mandatory tours are simulated.

Next, fully joint tours among household members are simulated, including who participates, the activity

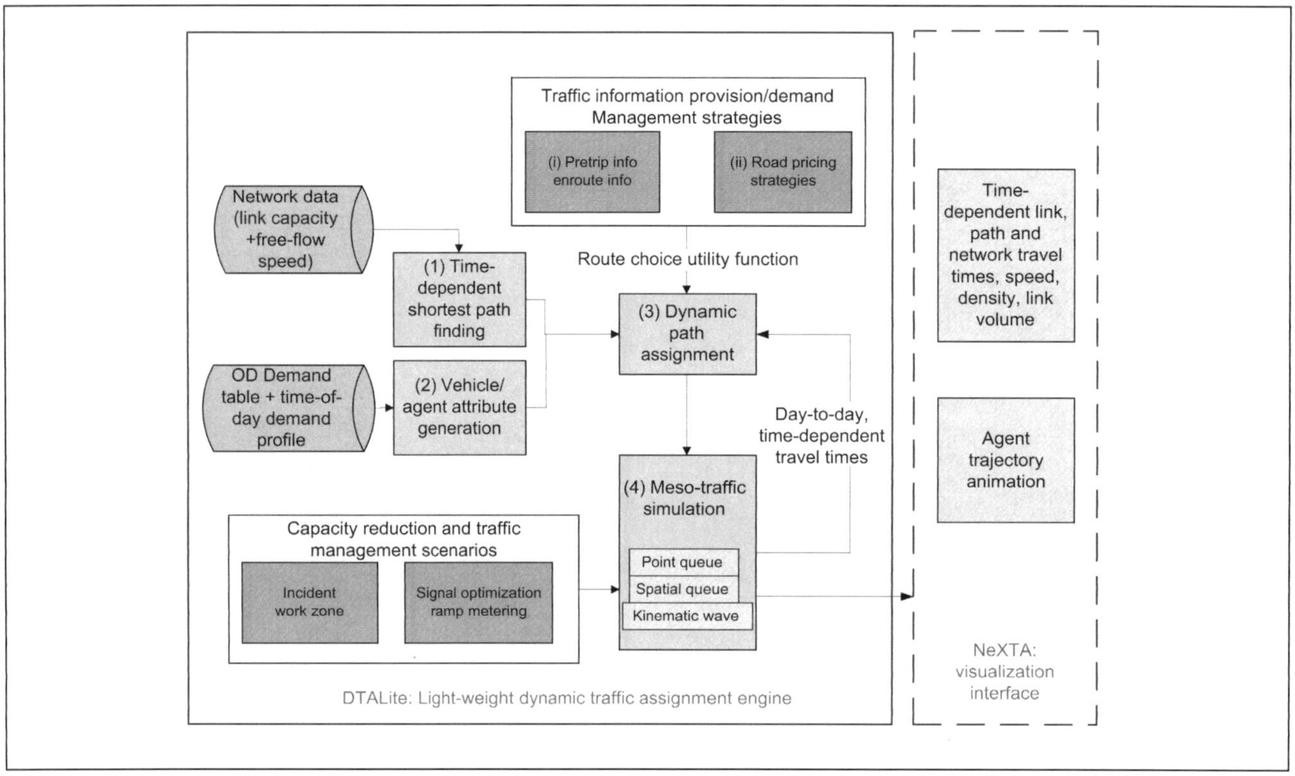

Figure 2. DTALite software system architecture with key modeling components.

purpose, the destination, and the time of day. After the details of the mandatory and joint tours are known, individual, non-mandatory tours are simulated, with their destinations and times of day. The final tour level models are the generation of stops for each tour and tour level mode choice.

The stop and trip level choices are simulated next. These include destination choice for each stop, the times for the stops, and mode choice for the trips between stops (conditional on tour mode). From the results of these models, auto vehicle and transit person trip tables are assembled for input to highway and transit assignment, respectively.

DTALite

DTALite is an open-source, light-weight, mesoscopic DTA simulation package; in conjunction with the Network eXplorer for Traffic Analysis (NeXTA) graphic user interface. It has been developed to provide transportation planners, engineers, and researchers with a theoretically rigorous and computationally efficient traffic network modeling tool. DTALite adopts a new software architecture and algorithm design to facilitate the most efficient use of emergent parallel (multi-core) processing techniques, and exploit the unprecedented parallel computing power newly available on both laptops and desktops.

The overall structure of DTALite, illustrated in Figure 2, integrates the four major modeling components, highlighted in yellow. DTALite's four major modeling components are: (a) time-dependent shortest path finding, based on a node-link network structure; (b) vehicle/agent attribute generation, which combines an origin–destination (OD) demand matrix with additional time-of-day departure time profile to generate trips. This component is only used to incorporate external OD demand in this project; (c) dynamic path assignment module, which considers major factors affecting agents' route choice or departure time choice behavior, such as (i) different types of traveler information supply strategies (e.g., historical, pre-trip and/or en-route information, and variable message signs), and (ii) road pricing strategies where economic values are converted to generalized travel time; and (d) a class of queue-based traffic flow models that can accept essential road capacity reduction or enhancement measures, such as work zones, incidents, and ramp meters.

The traffic assignment and simulation modules are fully integrated and iterated either to capture day-to-day user response or to find steady-state equilibrium conditions. Within this simulation–assignment framework, the

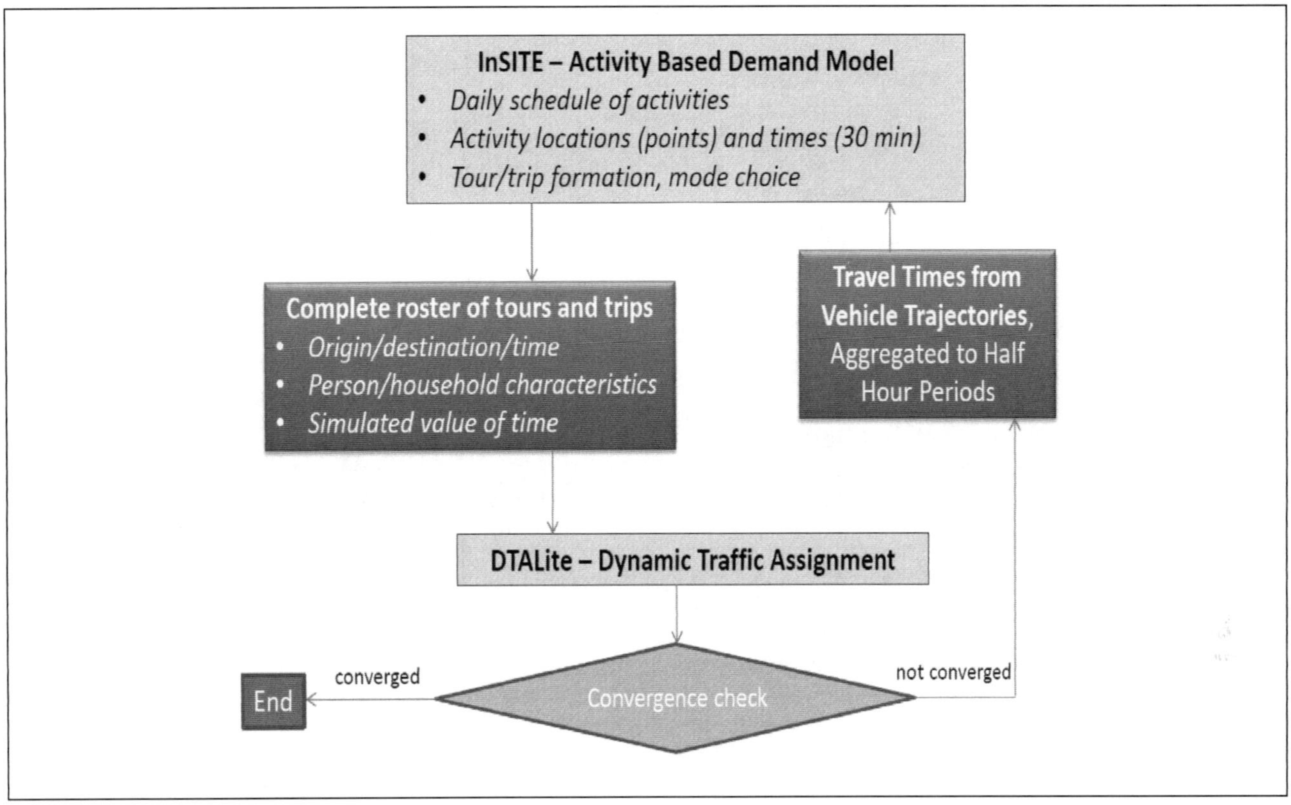

Figure 3. Sequential integration process.

rich set of output data includes traffic measures of effectiveness at different spatial and temporal scales, including network, corridor-level, and specific links. Typical speed, volume, and density measures, and agent-based trajectories, can be visualized through the NeXTA user interface.

Based on the design structure and queue-based mesoscopic traffic simulation model, DTALite has considerable potential for generalizing the modeling framework into the field of real-time traffic state estimation and prediction. In future research, different ways will be examined for calibrating the maximum queue discharge rates, utilizing end-to-end travel times, and considering an agent-based learning framework to consider behavioral heterogeneity more fully.

Integration

Overview of Integration Process

In this project, the sequential iteration process is followed. In this process, the ABM and the DTA model interface with each other, passing trip patterns and travel times back and forth until a measure of convergence is achieved. The general process is illustrated in Figure 3. The integrated model runs in iterations, with a single "big loop" consisting of an iteration of the travel demand model, followed by a run of the DTA model. The travel demand information (e.g., agents with their activities and travel decisions as well as characteristics) is passed to the DTA model within each big loop, and travel time information resulting from the dynamic assignment is passed back to the travel demand model for use as input (e.g., update activity pattern choices) in the next big loop. Note that DTA models themselves are run iteratively, and so there are several "small loops" of the DTA run within each "big loop." A number of "big loops" are run until a measure of convergence is achieved, in which the change in travel times (or some other measure) from one big loop to the next is within a specific tolerance. This methodology seeks practical convergence between the models since absolute convergence may not be achieved due to inherent differences at the various modeling levels. The data exchange between InSITE and DTALite is such that an agent is preserved during the exchange of data from InSITE to DTALite and vice versa. Thus, the integrated InSITE–DTALite model is fully disaggregated.

Software Implementation

In this study, intermediate modules are developed to facilitate the data exchange between InSITE and

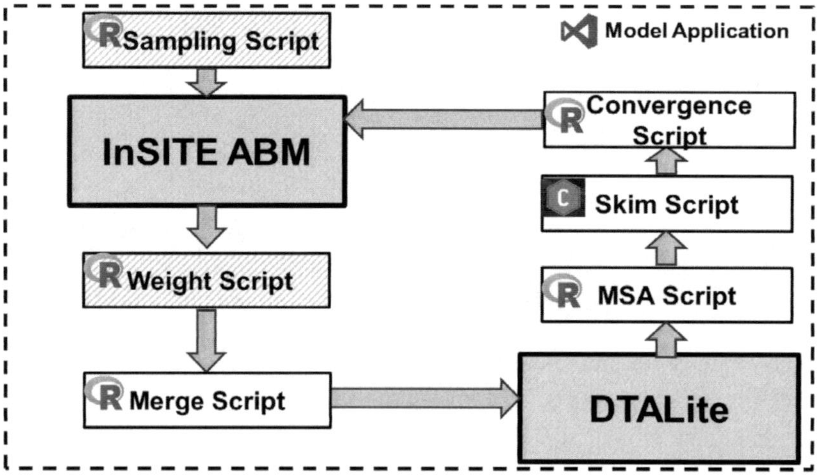

Figure 4. Software implementation (shaded sampling and weight scripts are optional).

DTALite. As illustrated in Figure 4, several R scripts are developed to change data formats and calculate convergence measures. The Sampling and Weight scripts (optional) are parts of the stratified sampling approach to improving model runtime, which is discussed in the Sampling Approach subsection. The Merge script is only executed after the initial big loop iteration to merge re-simulated agents, which is elaborated in the Convergence section, below. The MSA script employs the method of successive averages to calculate the average travel time of the previous iteration and the current iteration, which avoids overly rapid changes in network skims. The Skim script is written in CUBE software to convert DTALite network skims to a format that is compatible with InSITE. Finally, the Convergence script measures the activity schedule gap to determine if the convergence has been reached.

To execute various model components in one platform, a Visual Basic application is developed as a wrapper to call different modules. Users could customize the model run by specifying a simple configuration file to decide the number of big loop iteration or even to pick up a previously paused run. Since the integrated model run typically takes a long time to run, an e-mail notification function is also developed to inform the users of model status to avoid frequent status checking.

Information Exchange between Models

In the sequential integration paradigm, the ABM and the DTA model exchange information when the model run is complete. At the end of each InSITE run, the program will output a complete roster of trip information and traveler characteristics (e.g., OD, time of travel, value of time (VOT), etc.). One feature of InSITE is that each traveler in the population has a simulated VOT obtained from the VOT population density functions. These simulated VOTs are passed from InSITE to DTALite to use in its route choice models, which could also support the analysis of tolling scenarios. Compared with previous integration efforts, where the ABM outputs three separate files containing information regarding trip, person, and household, respectively, this DTA roster combines all information in one file, which improves the data exchange efficiency. This roster later becomes the input to DTALite so that the network representation can be simulated.

The main data passed from DTALite to InSITE is related to highway travel times. In the integrated model, DTALite simulates highway travel dynamically over the entire day from beginning to end. Since InSITE uses 48 aggregate time periods of 30 minutes in length, DTALite creates link travel times for each 30-minute period by averaging the times experienced by the vehicles (from trip trajectories) in the DTA during the period. It should be noted that, in the initial iteration, InSITE uses skimmed travel times for four aggregate time periods from a loaded network that has already reached equilibrium from a static assignment. After the initial iteration, the travel time inputs to InSITE can be unique for each half-hour period.

Sampling Approach

In the literature on ABM and DTA model integration, model runtime remains a major challenge since both demand and supply models employ a micro-simulation

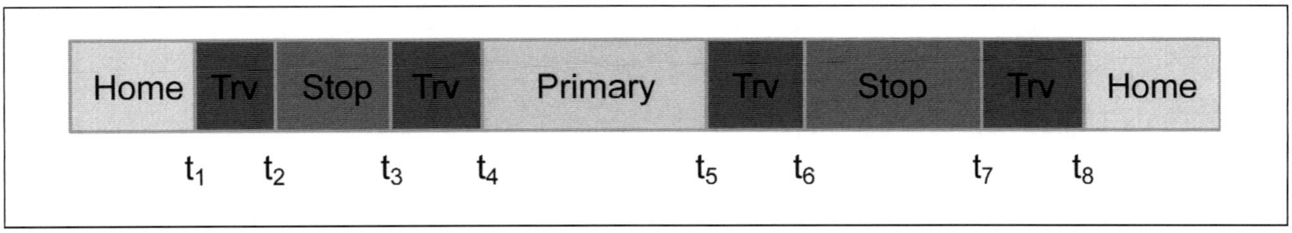

Figure 5. Activity schedule sequence example.

approach. In this study, InSITE takes up more than 50% of the total runtime. To decrease the InSITE model runtime, an alternative sampling approach is proposed, where InSITE first runs with a stratified population sample. A Weight script is later executed to expand the sample to full population, as illustrated in Figure 4. In this study, the population sample is stratified on two variables: home zone and income class. Two different population samples (10% and 25%) are tested in this study; they could decrease InSITE runtime by 50% and 27%, respectively.

The purpose of this stratified sampling example is to illustrate how much runtime this sampling approach could save. As for the representativeness of the population sample, it depends on what metrics the analyst cares about to stratify, and metrics for both the ABM and DTA model should be considered. In this study, we stratified the population based on the home zone and income because it is critical to cover all ODs or as many as possible and also to cover the VOT range since the VOT is used in the route choice function of DTALite and income is a proxy for VOT. Consequently, a good coverage of ODs and VOT could ensure that the travel times and traffic flows are comparable with those for the full population. In general, a represented sample is difficult to construct and therefore this approach is completely optional in running the integrated model.

Convergence

The convergence paradigm adopted for the sequential integration is based on the concept of activity-schedule gaps and is explained as follows: imagine you have an activity-schedule sequence for an agent, such as the one shown in Figure 5, where Primary represents the primary activity of the travel activity; Stop represents secondary activities, such as shopping and eating out; Trv represents traveling by a given mode (car, transit, etc.); and t_s represents departure times and arrival times of the activities.

Convergence is based on measuring scheduling gaps, which are defined as the changes in travel times between the previous skims and the current travel time skim for the same Trv element. Furthermore, agents traveling where the current travel time skims reduce their activity durations to less than the minimum activity durations allowed are identified. The minimum activity durations are obtained from the Household Travel Survey. These agents are flagged as infeasible and are re-simulated by InSITE ABM in the next big loop iteration. Agents not flagged as infeasible are not re-simulated by InSITE ABM in the next big loop iteration. The activity schedules of both agents re-simulated and agents not re-simulated are combined and provided to DTALite in the next big loop iteration to perform DTA, which is shown as the Merge script in the software implementation graph (Figure 4). It should be noted that an agent not re-simulated in one big loop iteration could be re-simulated in another big loop iteration. In general, the number of infeasible agents will decrease with probability one as more big loop iterations are completed, if the travel times converge with probability one to a constant value compatible with the minimum activity durations.

DTALite simulations across big loop iterations will become consistent as the number of infeasible agents decreases to less than 50,000 agents. In other words, link volumes and OD volumes generated by DTALite across big iterations will become approximately the same for all links. This is a result of DTALite being deterministic and only InSITE ABM being stochastic. The stochasticity in the travel decisions will not significantly affect DTALite, as only the travel decisions of the shrinking infeasible agents change across big loop iterations and the travel decisions of the feasible agents become constant across iterations. However, the number of infeasible agents is stochastic; it is likely that it will not reach zero and will likely fluctuate (increase and/or decrease across iterations eventually). DTALite results are still approximately consistent across big loop iterations if the number of infeasible agents is significantly small, compared with the total number of agents. The relationship between number of infeasible agents and number of big loop iterations is shown in Figure 6.

Validation

Validation is a critical element in ensuring that the developed model is robust and that agencies can use the finished product. Establishing a validation plan early in the

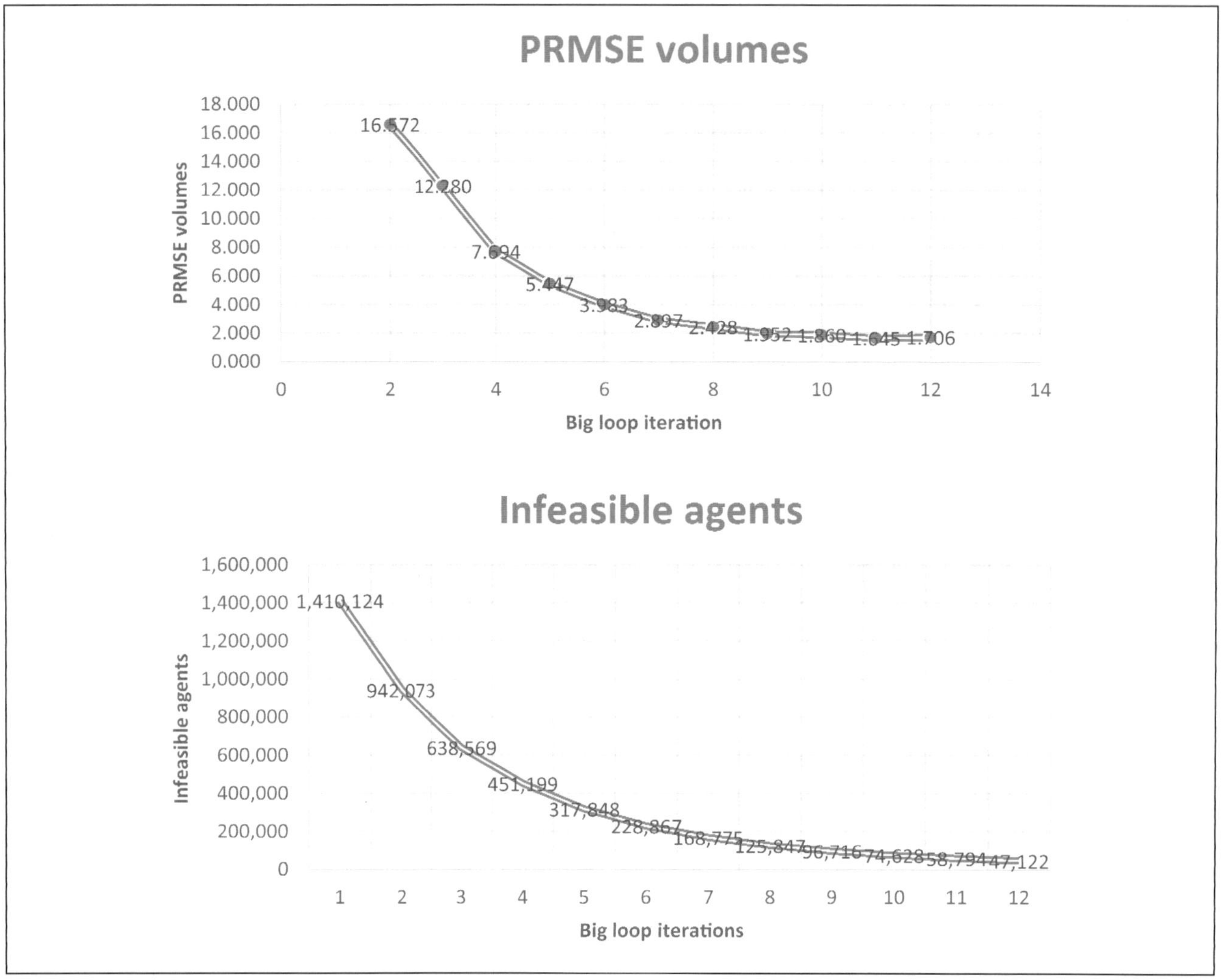

Figure 6. PRMSE volumes and infeasible agents versus big loop iterations.

model development process helps ensure that appropriate validation data are assembled or collected, and that adequate time and resources are devoted to the validation process. It should also be noted that the InSITE ABM (with static assignment) has been extensively validated against the Household Travel Survey data and traffic count data, and a number of sensitivity tests were also performed for InSITE. The validation of the integrated model focuses on DTALite trip assignment results. In this study, a structured, two-level approach was used:

- The Level 1 validation compares the traffic assignment results from InSITE ABM static assignment, with the DTA assignment results for volumes and vehicle-miles traveled (VMT).
- The Level 2 validation is comparable with traffic assignment validations typically performed for regional travel models. Generally, the Level 1 tests include aggregate (groups of links) comparisons of model volumes to observed data for the model base year.

The primary focus of the trip assignment validation was the ability of the integrated model assignment to reproduce observed daily traffic volumes and VMT. This validation can be considered a model system validation, since it is affected by the travel models embodied in both InSITE and DTALite. Figure 7 illustrates how the integrated model results and static assignment results compare with the observed traffic count data. In this figure, the X-axis and Y-axis denote the observed traffic counts and model results of a specific link, respectively, and each dot represents the link volume. Essentially, the closer one dot is to the diagonal line, the more accurate the model

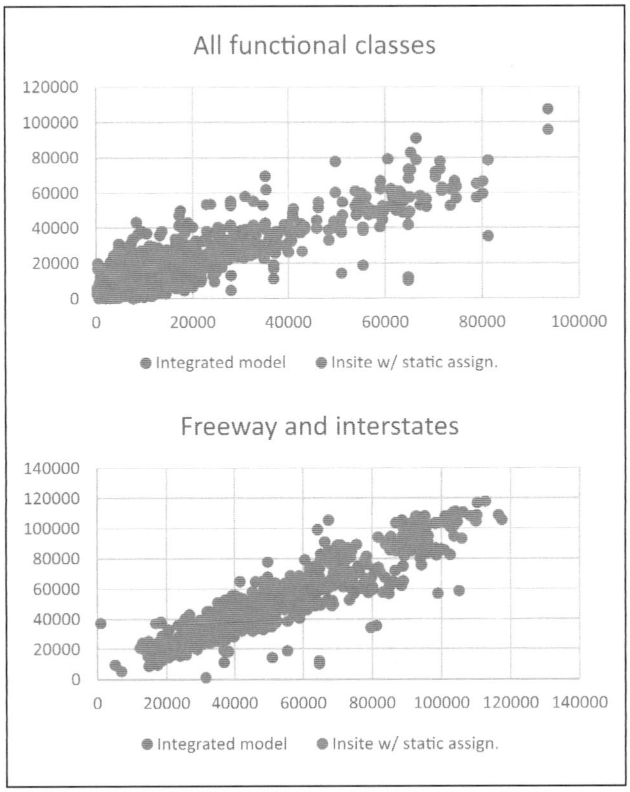

Figure 7. Observed daily volumes versus model daily volumes.

result is. The overall volume difference between the integrated model results and traffic counts is only 0.84%. Specifically, the integrated model does a good job of estimating traffic volumes for urban facilities with an error of 0.41% while the estimation error for the predominantly lower volume rural facilities is 12.54%. For different functional classes at an aggregate level, the overall validation errors are 3.69% for interstates, 3.24% for freeways, 5.78% for arterials, and 2.65% for collectors and other facilities, based on percentage root mean squared error (PRMSE) measures. In general, the results of the integrated model are more accurate than the static assignment results compared with the observed daily count data, even when looking at specific functional classes, with some exceptions. However, additional validation (e.g. hourly validation) is required to assess further the performance of the integrated model.

Model Application

The intent of this model application is to test the sensitivity of the integrated model using alternative input data and assumptions. Under Armour, an American sporting goods manufacturer, recently announced plans to develop about 50 acres south of Cromwell Street, Baltimore, as its Port Covington headquarters over the next 20 years. In response to this redevelopment, BMC has created a land-use change scenario in the brownfield area at Port Covington, with 2012 as the base year. In the land-use change scenario, 12,000 new residents and 14,000 new jobs are introduced to two traffic analysis zones (TAZs) near Port Covington. With the integrated model, the impacts of this redevelopment can be analyzed. The results include the finding that the total traffic volume and travel time of all the links in these two TAZs will increase by 22.63% and 11.31%, respectively. In addition to these aggregated measures, the integrated model is able to analyze the traffic conditions at a much finer time resolution due to the treatment of time as a continuum. For instance, the 24-hour volume and travel time increase of the proposed scenario compared with the baseline scenario is illustrated in Figure 8. As mentioned above, both the demand (ABM) and supply (DTA) models incorporate the half-hour interval temporal representation, which makes it possible to represent the network dynamics at a shorter time interval. The proposed model application is only one example of the expanded analysis capabilities of the integrated model. Other popular active transportation and demand management strategies (e.g., dynamic speed limits, ramp metering, dynamic speed limits, etc.) can also be analyzed with the proposed integrated modeling system.

Conclusions and Future Research

In summary, an integrated model composed of the InSITE ABM and the mesoscopic traffic simulator DTALite is developed, which covers most of the urbanized areas in Maryland. The integrated model will support long-range planning, transportation conformity analysis, investment programming, project prioritization, equity analysis, and corridor/project planning at both SHA and BMC. The integrated model has been peer reviewed by the Technology Readiness Level panel from SHRP2 C10 program and graded as implementation-ready, which implies that it has been validated and includes all functionality required for typical planning analyses. The validation results suggest that the integrated model outperforms the static assignment-based InSITE model. The integrated model is also applied to analyze a land-use change scenario in Port Covington, Baltimore. The treatment of time as a continuum enables the integrated model to present results at finer time resolutions, which could better support decision making. The contributions of this study can be summarized by the following four aspects: (a) a stratified sampling approach is developed to decrease the model runtime; (b) new convergence criteria based on agents with infeasible schedules are proposed; (c) efficient data exchange through

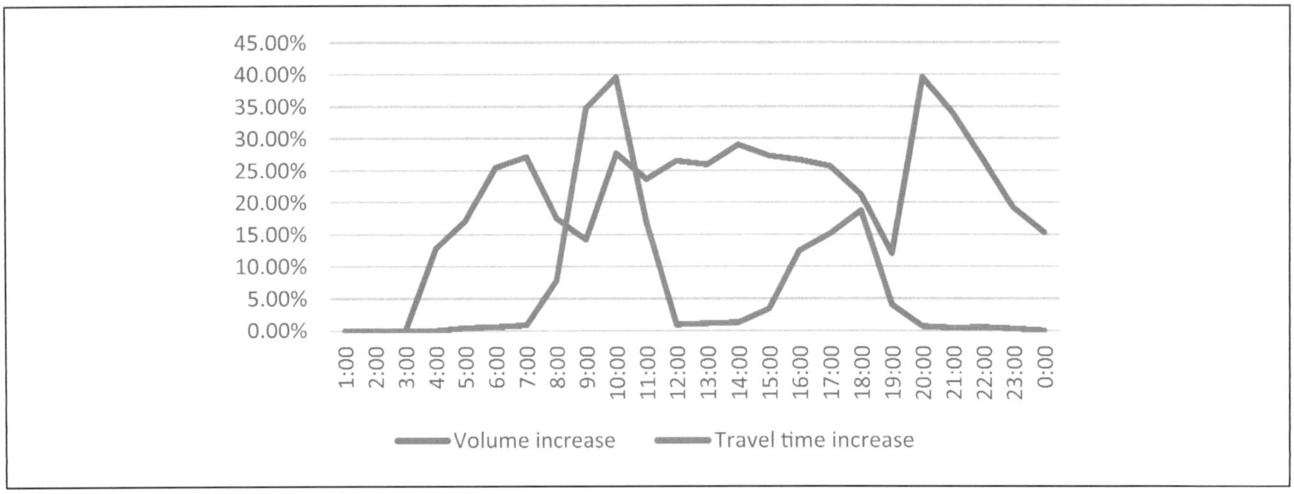

Figure 8. Volume and travel time increase compared with baseline scenario.

one single trip roster file; and (d) implementation experience in the Baltimore-Washington metropolitan region is valuable to both researchers and practitioners.

To further extend this study, two future research directions are proposed in this paper. One research direction is to expand the integrated model system to the statewide level. The current Maryland statewide transportation model is a trip-based, four-step model. A statewide ABM and DTA integrated model would enable policy makers to observe the effects of their policies on different segments of population (e.g., income groups), which will lead to more equitable policies. In addition, it will help monitor how travelers' behavior and their activities change after certain transportation or land-use policy is implemented. Furthermore, the DTA capabilities will provide an analytical engine to understand non-recurrent congestion management strategies at the statewide level (e.g., incident management, work zones management, emergency planning, special events planning, and others).

The integration approach introduced in this study falls in the sequential integration paradigm. As simple as this integration approach is, it is not responsive to network dynamics or real-time information technologies. A more realistic integration approach is the dynamic integration where the ABM and the DTA model communicate in a much more frequent manner. Another future research direction of this study is to pursue a dynamic integration between InSITE and DTALite. Considering that InSITE and DTALite will run in parallel in a dynamic integration, a more efficient data exchange method than the simple file reading and writing is required to control the model runtime within a reasonable time frame. Meanwhile, it is also difficult to run both InSITE and DTALite at the same time in terms of computational resource allocation. While challenging, the dynamic integration between InSITE and DTALite is the primary next step for this study, as it allows for modeling dynamic mobility management strategies as the profession seeks to integrate planning and operations through the integrated model system.

Acknowledgments

This research is financially supported by the U.S. Federal Highway Administration Strategic Highway Research Program 2 (SHRP 2) project, Maryland State Highway Administration, and Baltimore Metropolitan Council.

Author Contributions

The authors confirm the contribution to the paper as follows: study concept and design: LZ, DY, SG, CC, CX, TFR, SM and CB; integration method proposal and implementation: DY, SG and CC; BMC InSITE and integration technical support: TFR and MM; result analysis and interpretation: LZ, DY, SG, CC, SM and CB; manuscript preparation: LZ, DY, SG and CC. All authors reviewed the results and approved the final version of the manuscript.

References

1. *SHRP 2 Report S2-C46-RR-1: Activity-Based Travel Demand Models: A Primer*. Transportation Research Board of the National Academies, Washington, D.C., 2015.
2. Lin, D. Y., N. Eluru, S. Waller, and C. Bhat. Integration of Activity-Based Modeling and Dynamic Traffic Assignment. *Transportation Research Record: Journal of the Transportation Research Board*, 2008. 2076: 52–61.
3. *Transportation Research Circular E-C153: Dynamic Traffic Assignment: A Primer*. Transportation Research Board of the National Academies, Washington, D.C., 2011. http://onlinepubs.trb.org/Onlinepubs/circulars/ec153.pdf.

4. Pendyala, R., D. You, V. Garikapati, K. Konduri, and X. Zhou. Paradigms for Integrated Modeling of Activity-Travel Demand and Network Dynamics in an Era of Dynamic Mobility Management. Presented at 96th Annual Meeting of the Transportation Research Board, Washington, D.C., 2017.
5. Balmer, M., M. Rieser, K. Meister, D. Charypar, N. Lefebvre, and K. Nagel. MATSim-T: Architecture and Simulation Times. In *Multi-Agent Systems for Traffic and Transportation Engineering*, IGI Global, Hershey, PA, 2009, pp. 57–78.
6. Pendyala, R., K. Konduri, Y. C. Chiu, M. Hickman, H. Noh, P. Waddell, and B. Gardner. Integrated Land Use-Transport Model System with Dynamic Time-Dependent Activity-Travel Micro-Simulation. *Transportation Research Record: Journal of the Transportation Research Board*, 2012. 2303: 19–27.
7. Auld, J., M. Hope, H. Ley, V. Sokolov, B. Xu, and K. Zhang. POLARIS: Agent-Based Modeling Framework Development and Implementation for Integrated Travel Demand and Network and Operations Simulations. *Transportation Research Part C: Emerging Technologies*, Vol. 64, 2016, pp. 101–116.
8. Lin, D. Y., N. Eluru, S. Waller, and C. Bhat. Integration of Activity-Based Modeling and Dynamic Traffic Assignment. *Transportation Research Record: Journal of the Transportation Research Board*, 2008. 2076: 52–61.
9. Bekhor, S., C. Dobler, and K. Axhausen. Integration of Activity-Based and Agent-Based Models: Case of Tel Aviv, Israel. *Transportation Research Record: Journal of the Transportation Research Board*, 2011. 2255: 38–47.
10. Goulias, K. G., C. R. Bhat, R. M. Pendyala, Y. Chen, R. Paleti, K. C. Konduri, and H. H. Hu. Simulator of Activities, Greenhouse Emissions, Networks, and Travel (SimAGENT) in Southern California. Presented at 91st Annual Meeting of the Transportation Research Board, Washington, D.C., 2012.
11. RSG, AECOM, Mark Bradley Research and Consulting, John Bowman Research and Consulting, Hadi, M., R. Pendyala, C. Bhat, and T. Waller, and North Florida Transportation Planning Organization. *Dynamic, Integrated Model System: Jacksonville-Area Application*. The National Academies Press, Washington, D.C., 2014.
12. Cambridge Systematics, Inc., Sacramento Area Council of Governments, University of Arizona, University of Illinois, Chicago, Sonoma Technology, Inc., and Fehr and Peers. *Dynamic, Integrated Model System: Sacramento-Area Application*. The National Academies Press, Washington, D.C., 2014.

The Standing Committee on Metropolitan Policy, Planning, and Processes (ADA20) peer-reviewed this paper (18-03713).

The opinions, findings, and conclusions or recommendations expressed in this paper are those of the authors and do not necessarily reflect the views of the U.S. Federal Highway Administration, Maryland State Highway Administration, or Baltimore Metropolitan Council.

Article

TRR
JOURNAL OF THE TRANSPORTATION RESEARCH BOARD

Transportation Research Record
2018, Vol. 2672(51) 56–67

DOI: 10.1177/0361198118796019
journals.sagepub.com/home/trr
SAGE

Are Small- and Medium-Sized Community Paratransit Riders Ready to Adopt Real-Time Information (RTI) Technology?

Jeffrey J. LaMondia[1], Travis Gajkowski[1], and Veronica Ramirez[1]

Abstract

This research seeks to understand the factors affecting who, in small and medium-sized communities, would adopt real-time information (RTI) technology when using paratransit services. It further provides a planning performance measure that can identify areas which would benefit most from the introduction of such RTI technology. These goals are addressed by modeling paratransit riders' likelihood of adopting RTI technology as well as forecasting communities' likely adoption using on-board passenger survey data collected in two small- and medium-sized communities in Alabama and Georgia. Overall, the estimation parameters highlight that future generations of paratransit patrons who are familiar with general technology and use paratransit for non-routine trips would be most interested in and gain the most benefit from this RTI technology. The performance measure is also most effective in long-term transportation planning, applied to future older populations (currently 45–55 years old) who are more comfortable using technology. Transit agencies will be able to use this information to determine whether RTI would be appropriate for their communities as well as the populations that should be targeted for this introduction prior to undertaking a large study or infrastructure investment.

Paratransit, sometimes referred to as demand-response transit (DRT) or dial-a-ride transit, is a critical mode for small and medium-sized communities with widely-dispersed populations that cannot support a fixed-route service (*1*). While paratransit is often associated with mobility programs under the Americans with Disabilities Act of 1990 (ADA), which are required to support citizens with disabilities and mobility challenges, there is a significant demand for non-ADA dial-a-ride transit services, most often centered on older populations (*2*). In fact, APTA estimates that ridership of non-ADA dial-a-ride transit services will continue to expand by up to 80% over the next twenty years (*3*). This is driven by the aging U.S. population, which will see more than 20% of the population aged 65 or older by 2030 (*4*). As such, transit planners and engineers are seeking efficient means for supporting and expanding paratransit services to provide sustainable and equitable transportation options for the increasingly aging citizen demand.

Unfortunately, providing this support can be challenging, as paratransit operations can be costly and complex. When compared with a fixed-route transit service, paratransit costs per trip can be up to 10 times more expensive. However, the user cost associated with paratransit services is competitive with or even lower than general public transportation. The additional costs associated with paratransit can require specialized vehicles, such as lift-equipped cutaway buses that serve physically disabled passengers. Paratransit operators also require specialized training in order to assist the needs of their passengers such as wheel chair securement (*5*).

One potential cost-effective solution is the use of advanced communication and real-time information (RTI) technology. Over the past decade, many transit agencies across the country have introduced websites, smartphone applications, and informational displays that provide riders with real-time travel updates, travel planning tools, and the ability to book a trip from home. Paratransit and general transit riders value knowing the length of their waiting time, allowing them to plan accordingly. These improvements have resulted in increased general levels of customer satisfaction (*6*), increased the paratransit system's reliability in delivering

[1]Department of Civil Engineering, Auburn University, Auburn, AL

Corresponding Author:
Address correspondence to Jeffrey J. LaMondia: jlamondia@auburn.edu

accurate waiting times, and in some cases increased passenger ridership (*7*, *8*).

Despite the success of these technologies for fixed-route transit systems, it is relatively uncommon for this technology to be applied to paratransit services. Online scheduling services can be used in paratransit operations (*9*), which could be integrated with mobile or web-based communications to inform riders about up-to-date schedules, arrivals, and service alerts. Consequently, there are ample opportunities to improve the mobility and access of paratransit users through the use of RTI, and this research seeks (a) to understand the factors affecting who would adopt RTI paratransit technology and (b) to provide a planning performance measure that can identify which areas would benefit most from the introduction of such RTI technology.

These goals are addressed by modeling paratransit riders' likelihood of adopting RTI technology as well as assessing the benefit communities would see from implementing this technology using data collected in two small and medium-sized communities in Alabama and Georgia. Transit agencies will be able to use this information to determine whether RTI would be appropriate for their communities as well as the populations that should be targeted for this introduction. The paper includes the following sections: a discussion of the role of RTI transit technology, a description of the on-board transit survey collection, an evaluation of the interest in RTI based on rider demographics, an ordered probit regression model of rider adoption, a summary of the adoption likelihood performance measure, and conclusions.

Role of RTI Technology in Transit Systems

There are four main types of RTI transit technology: real-time arrival prediction software, automatic vehicle location services, two-way SMS/text messaging systems, and alert subscription services. Many RTI applications use several or all of these RTI technologies simultaneously. The Transit Cooperative Research Program (TCRP) Synthesis 91 surveyed 28 of the largest transit agencies around the world in 2010 to determine what types of RTI technology the agencies employed and how they are used. Some examples of the variety of information RTI transit services provide include real-time vehicle location information from GPS, announcements (service disruptions, detours, emergencies, etc.), current route/destination information, and the most common RTI application—the time or distance until the next bus arrival/departure. There are many services available to transit agencies to provide RTI to their passengers. According to the 2010 survey conducted by TCRP, the most common mobile RTI service utilized by transit agencies mobile Internet services, followed by a smartphone app, mobile alerts, and two-way SMS/texting, with few agencies employing other methods such as near field communication or location-based services (*10*).

The most common application of RTI technologies is to predict bus arrival/departure times. For example, Bus Time in New York City has multiple access methods, including desktop and mobile websites and SMS (*7*). The Chicago Transit Authority uses Bus Tracker, a system that uses GPS devices to track real-time bus locations. The information is available for public use and multiple smartphone apps have been developed that can push real-time alerts to users about upcoming bus arrivals (*8*). However, not all RTI technology applications are mobile applications. Some transit agencies use RTI to update dynamic message systems (DMS) at bus stops, showing up-to-date arrival times (*11*). However, DMS usage is declining due to the cost of building and maintaining that infrastructure (*10*). Trip planning software integrated with RTI is another important use of these technologies that allows users to prepare their trip with the latest data available (*10*, *12*). Trip Planner is an example of one such program. With Trip Planner, users can obtain information about their origin and destination stops and real-time bus locations to plan their trip online or via SMS (*12*).

The main benefits of RTI include: decreases in both wait times and customer perception of wait times, increased satisfaction with the transit service, and increased ridership and transfers. A case study of bus riders in Tampa, Florida, examined how implementing a RTI system impacted bus riders. The research team found that the primary benefit to bus passengers in a RTI system was reduced passenger waiting times. Researchers determined that the average reduction in waiting time was nearly 2 minutes. Results from case studies in Stockholm, Sweden (*13*), King County, Washington (*14*), and New York City (*7*) also indicated that reduced waiting time was a critical benefit to passengers. The study in Stockholm determined that the benefits to users in wait times corresponded to a 60% increase in service frequency (*13*). Results from the case study in King County also indicated that passenger wait time was reduced by almost 2 minutes, or 30%, with RTI. In addition to reducing passengers' actual waiting time, the research team also determined that passengers without RTI perceived their wait time as longer than it actually was, a characteristic not shared by those who accessed the up-to-date scheduling information (*14*). Many researchers agree that developing an application for mobile devices is also a relatively inexpensive way to improve the transit level of service (*14*, *15*). Researchers from Chicago found that implementing a RTI service resulted in modest ridership gains of about 2% with relatively little investment (*8*). The study from New York

City followed a similar methodology with a very similar result of about 1.7% increase in ridership (*7*). According to a survey conducted in Thessaloniki, Greece, passenger satisfaction with the transit system rose by almost 8% after implementation of a RTI system (*11*).

Information on who are the users of RTI technologies on transit is scarce. A report conducted by Bregman et al. surveyed 130 transit agencies in the United States and Canada to ask them about their recent estimates of riders with access to technology. They found that many agencies had no estimates for this and more estimated that their ridership's access was lower than the national average. Their results indicate that an average of 65% of transit riders have access to Internet, compared with the national average of 85% (*16*). A unique study of the metro in Saint Louis, Missouri found that metro riders who used smartphone reported higher levels of satisfaction than non-users and that riders older than 40 years of age were less likely to use a smartphone (*17*). According to a study by the Pew Research Group in 2013, the elderly and those with low incomes are less likely to use the Internet. Of those aged 65 or older, 44% do not go online, compared with only 2% of those aged 18 to 29. As to socioeconomic factors, 24% of people making less than US$30,000 annually do not go online, while for those making US$75,000 or more a year, only 4% do not browse the web (*18*).

While these types of mobile applications are becoming more common in public transit, few paratransit facilities have implemented RTI technologies. Thus, little research exists on the impact of these applications to paratransit. A study of paratransit services in North America, Europe, and Asia determined that a mobile application that provides similar RTI and trip scheduling options as many transit apps could be the key to providing large-scale flexible paratransit options (*19*). A study by Davison et al. determined that, while current markets favor the dial-a-ride approach to scheduling paratransit trips, using technology to communicate more effectively with the customer base should be a priority in the future in order to attract a wider market (*20*). While paratransit is one of the most efficient ways to provide mobility to the elderly, those with disabilities, and those with limited access, it is also one of the most expensive transit options available. There are many challenges to operating an effective DRT system, including scheduling pick-ups and drop-offs, determining bus routes, informing customers of delays or other events, and coordinating the system to run efficiently (*21*). The best RTI technology for paratransit would include a trip planner that allows patrons to plan their trips online, via SMS, or with a smartphone app. This is a high priority, as it gives customers more options to schedule paratransit trips. Real-time predicted arrival/departure times could lower customer waiting times and allow them to use their time more efficiently. An app with alerts could serve a similar function, notifying customers when their bus is near.

Developing and implementing mobile RTI technologies also helps transit agencies communicate more effectively by increasing their information equity. The same is true for paratransit. Today, transit agencies face mounting public pressure for faster, easier access to information. By providing this information in a variety of forms, more people would have access to it and therefore, using paratransit would be easier for them. While much of the current transit ridership population has limited access to technological devices, providing more information to riders can increase the effectiveness of paratransit for all (*10*).

On-Board Survey Collection

On-board passenger surveys were collected on paratransit routes operated by Lee-Russell Public Transit (LRPT), which provides DRT services to some of the fastest growing small and medium-sized communities in Alabama. LRPT offers a dial-a-ride service across Lee and Russell counties, accommodating both ADA and non-ADA trips, Monday through Friday between 8:00 a.m. and 2:00 p.m. Passengers are required to phone in their trip requests at least 24 hours prior, and routine trip services are accommodated. LRPT's use of technology is limited, with no online scheduling service or smart device integration (*22*). Passengers on the LRPT service were selected to be surveyed because its paratransit services are open to all of the general public, including seniors and disabled passengers.

The single-page paper survey was collected in-person over the course of 4 days in April and June, 2016, while riding the vehicles, in an attempt to capture a wide variety of travelers, trip purposes, and experiences. The survey included questions about how often the respondents used LRPT services, whether they used it for different purposes (e.g., medical, shopping, religious, recreation, work/errands), how early they typically scheduled the trip, access to computer/ phone technology, and demographics (e.g., ethnicity, income, age, and gender). The topic of this research was captured in the following question: "Many paratransit services are moving to an online scheduling service. This allows you to: schedule your trip, see the location of your bus and see up-to-minute departure/arrival times. If one were available, how likely would you be to use an online scheduling service?" Respondents could choose Very Likely, Somewhat Likely, or Not Likely. Most respondents understood (either from the survey or from prior knowledge) what this technology was but mainly those aged 65 or older were unfamiliar with these topics and required the authors to provide more details about RTI and online scheduling services. In fact, many of them shared their thoughts on the technology; one passenger expressed their discontent with the extensive waiting time over the phone in order to request the service and mentioned that adaptation to this technology would be beneficial for the agency and the passengers. The final dataset includes 56 complete survey responses.

The survey response rate was nearly 100% (with only one decline) due to the fact that surveys were collected similar to one-on-one interviews. In most cases, the bus drivers introduced the survey and the researcher to each participant as he/she boarded, which established credibility and trust. Participants were assured they could stop at any time. Each researcher then asked participants whether they would be more comfortable completing the survey themselves or if they would like to complete it verbally; all but one completed the survey verbally. The researchers were friendly and talkative before and after the survey, but they became more objective while administering the survey questions.

Survey respondents are consistent with the LRPT ridership: 25.0% Caucasian and 57.0% African American; 45.5% male and 54.5% female; 17.9% aged 35 to 49, 32.1% aged 50 to 65, and 50.0% aged 66 or older; and 72.5% with a household income less than US$25,000 and 27.5% with a household income of US$25,000 or higher. Additionally, LRPT supports many routine trips, scheduled on a recurring time and destination (e.g., dialysis or doctor appointments), as well as short-term trips, scheduled the day before. The survey also found that 50.0% of travelers scheduled their trip more than a week prior, 16.1% less than a week before, and 33.9% only one day before. Overall, the LRPT survey respondents (and ridership in general) are skewed to older populations with lower incomes and racial diversity.

Figure 1 shows a summary of different survey responses related to frequency of using paratransit,

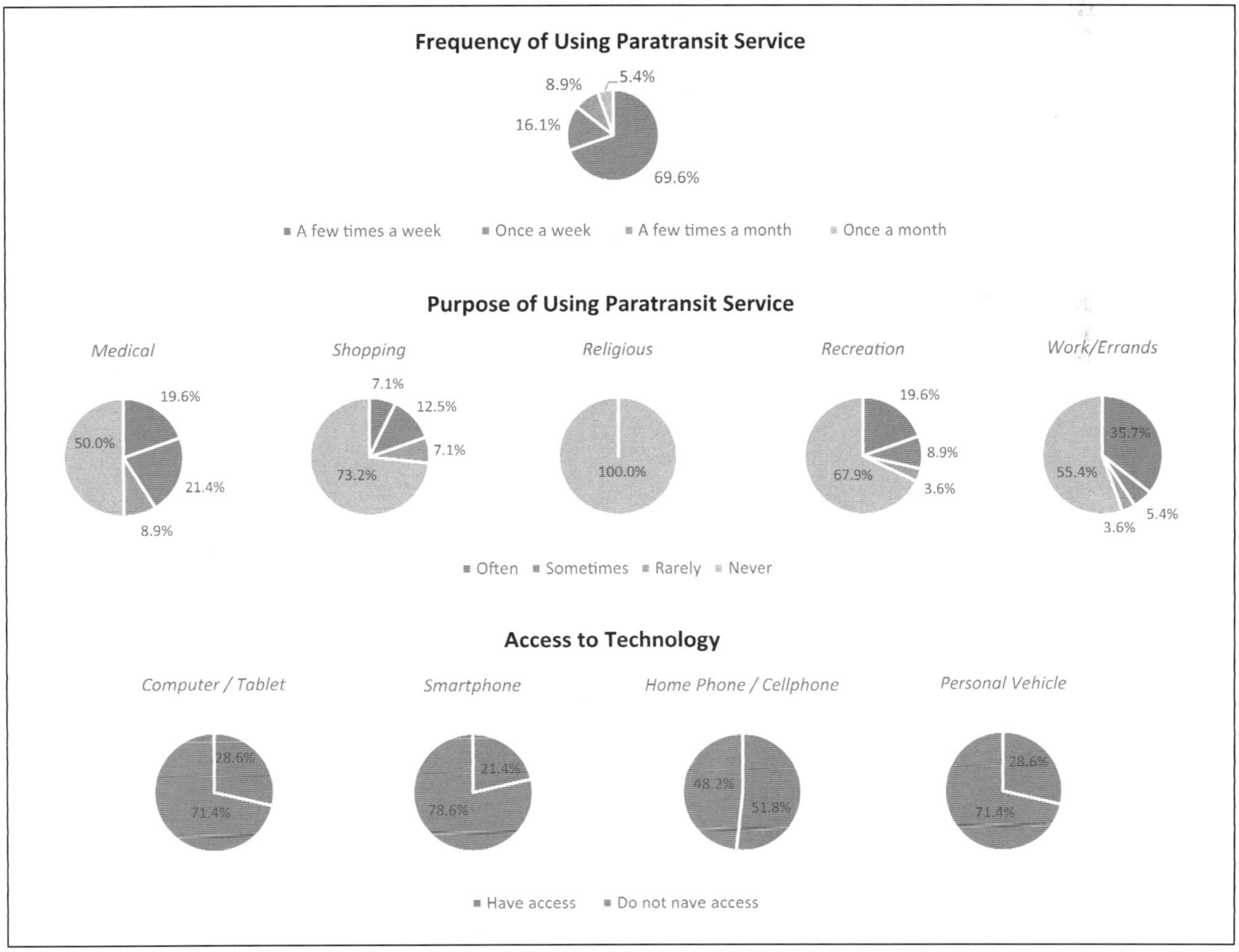

Figure 1. Survey response summaries.

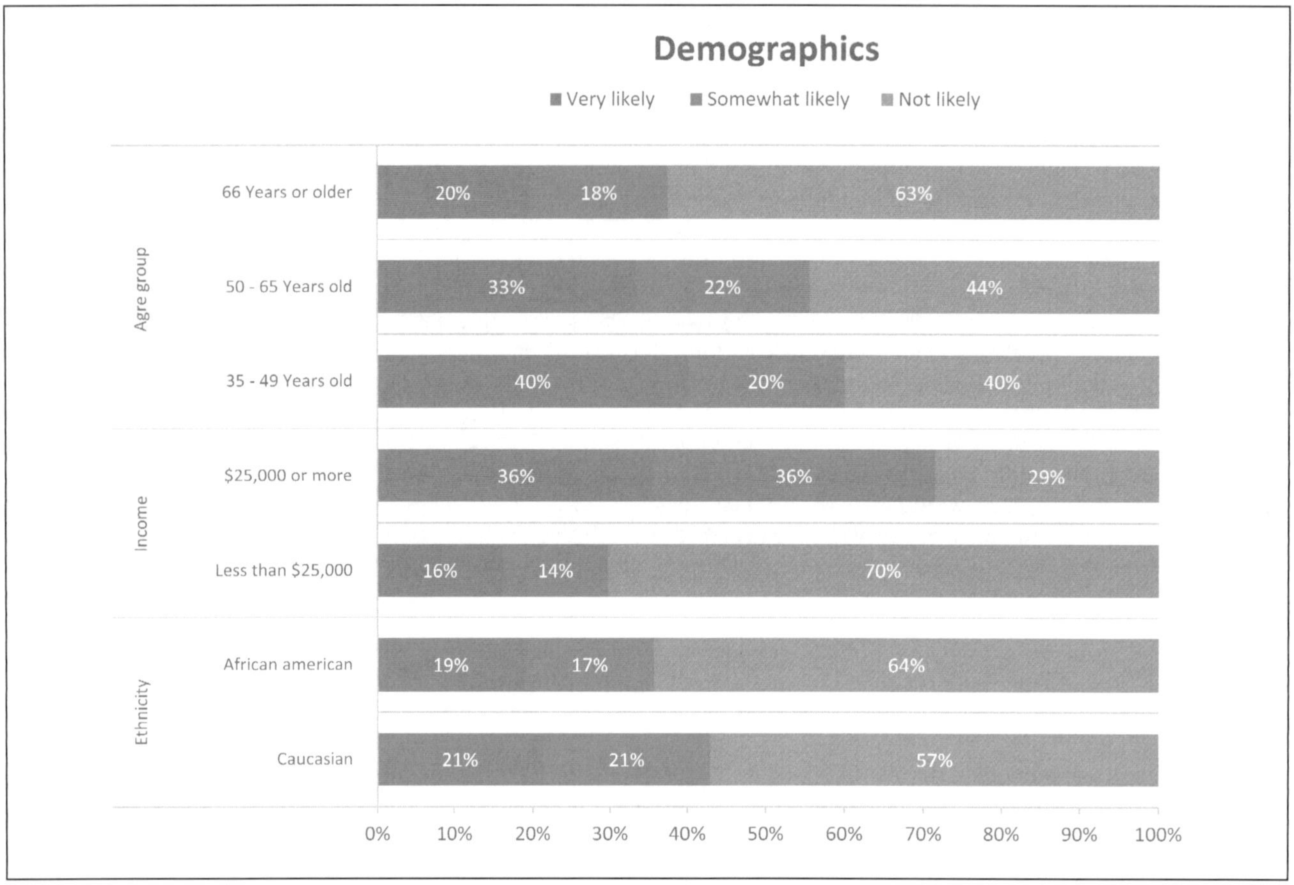

Figure 2. Respondents' likelihood of using RTI relative to demographics.

purposes for using transit, and access to technology. Most LRPT riders use the service frequently and depend on it to get around, as 71.4% do not own an automobile. Riders vary in their reasons for using the service, but most use it for medical purposes, commuting to work or running errands, or for recreational purposes, with few riders using the bus to shop and none using the service for religious travel. Very few have access to smart devices and nearly half do not own any type of telephone. Of those older than 65 years, 21.4% had access to a computer, 10.7% had access to a smartphone and 25.0% had access to a personal vehicle. Of those younger than 65, 35.7% had access to a computer, 32.1% had access to a smartphone and 32.1% had access to a personal vehicle.

General Interest in RTI Paratransit Technology

While the broad statistics seem to suggest that there is mixed interest in RTI technology for paratransit riders (62.5% stated they would be unlikely to use it, 17.9% stated they would be somewhat interested, and 19.6% stated they would be very interested), there are meaningful variations in the responses to RTI based on demographics, access to technology, and paratransit use. These distributions relative to interests are presented in Figures 2–4 and discussed below.

Figure 2 shows interest in RTI adoption based on demographics. One can see that as riders get older, they are dramatically less interested in using RTI technology. In fact, 80% of riders aged 35 to 49 years are either very or somewhat likely to adopt RTI technology, compared with just 38% of riders aged 66 or older. This is perhaps due to younger users being more familiar with technology in general. Not surprisingly, individuals with higher incomes (who may also be more familiar with technology) are significantly more likely to adopt RTI technology: 71% of individuals with income of US$25,000 or more state that they are very or somewhat likely to use RTI technology compared with just 30% of people with income less than US$25,000. There is no observed difference in adoption preferences across ethnicities.

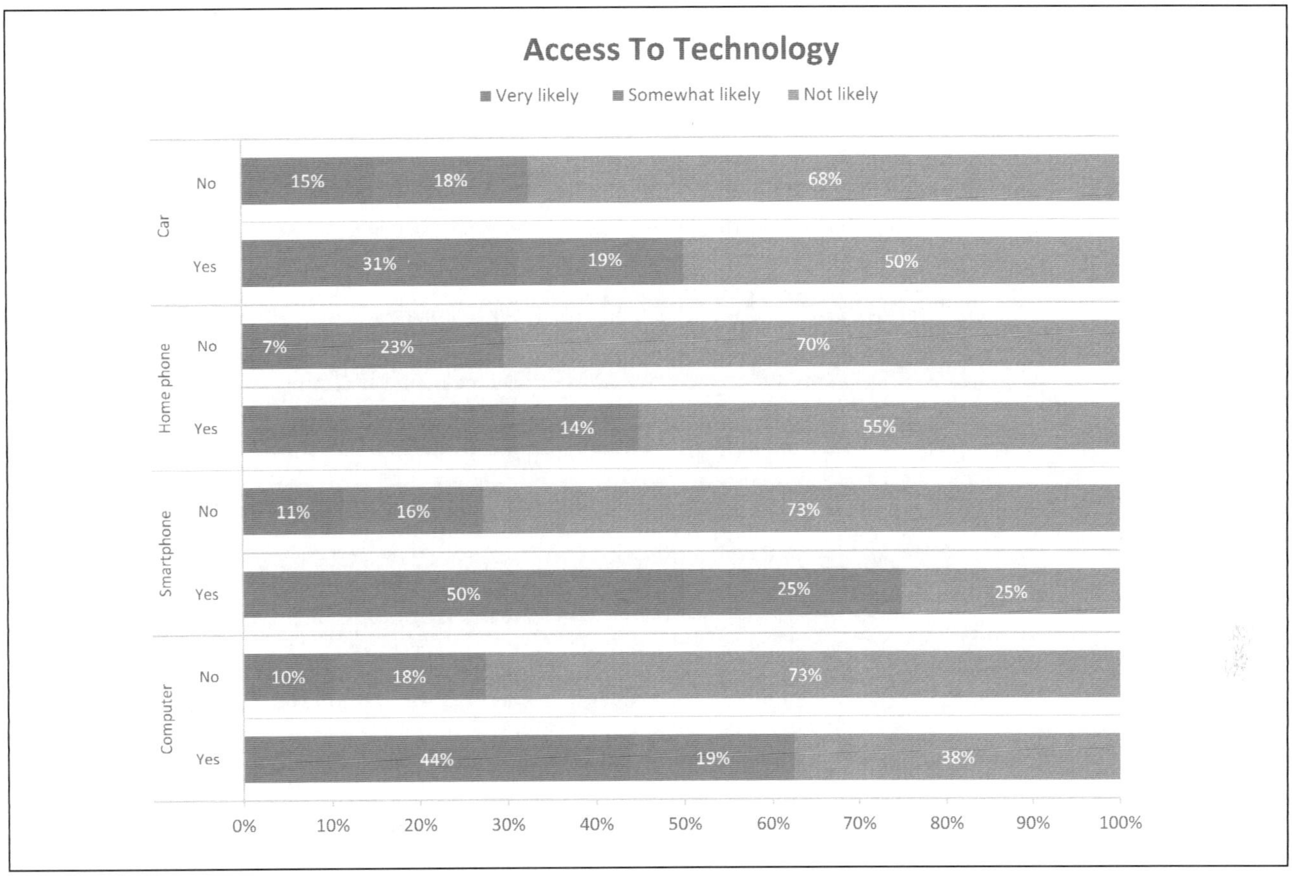

Figure 3. Respondents' likelihood of using RTI relative to access to devices.

Figure 3 shows interest in RTI adoption based on access to technology. The bar charts highlight that if travelers have access to technology they are significantly more interested in RTI technology. This makes sense, as those with experience using technology can appreciate the benefits and accessibility improvements they will gain by using RTI. The most extreme differences can be seen with smartphones and computer access. Those with smartphones are 2.78 times more interested in using RTI technology, and those with computers are 2.25 times more interested.

Finally, Figure 4 shows interest in RTI adoption based on how riders use paratransit. There is much more variability in these bar charts. Individuals traveling sometimes or rarely for each purpose have greater interest, being very or somewhat likely to adopt RTI, and no individuals who traveled for recreation or shopping would be very interested in using RTI. Interestingly, those respondents who travel often on the more mandatory activities of work and medical travel have some of the least interest in using RTI technology. Perhaps this indicates these repeat trips are so routine that technology would not offer additional benefits.

Ordered PROBIT Regression of Rider Adoption of RTI Technology

Next, an ordered logistic regression was estimated to understand, of all the factors identified in the summary statistics that affect adoption, which has the most relative significant impact on RTI adoption. The results indicate which characteristics (or combinations of characteristics) are most influential in RTI adoption and can be used for planning purposes.

The logistic regression compared many independent variable factors (seen in Table 1) by estimating weights for each, along with outcome thresholds, to accurately predict an ordered categorical dependent choice variable—likelihood of using RTI technology—with three alternatives: very likely, somewhat likely, and not likely. The model assumes there is an unobserved continuous underlying latent function (Y_i^*) approximating individuals' interest in RTI technology written as

$$Y_i^* = \sum_{n=1}^{N} \beta_n x_{ni} + \varepsilon_i \tag{1}$$

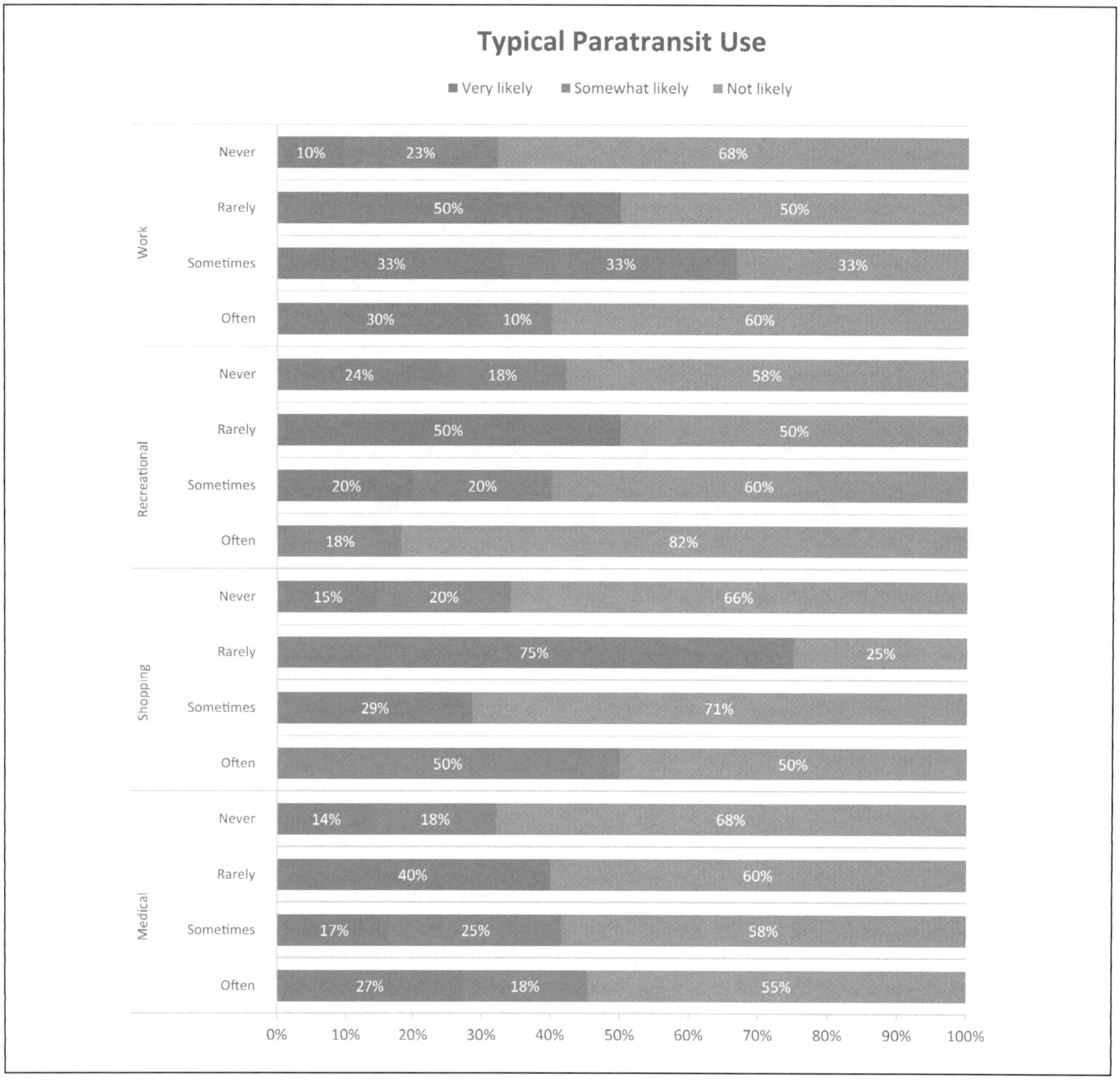

Figure 4. Respondents' likelihood of using RTI relative to typical paratransit use.

where I is the number of individuals, N is the number of independent variables, x_{ni} is the value of variable n for individual i, β_n is the coefficient weight of variable n, and ε_i is a random error term assumed to be normally distributed. The actual choice of using RTI technology (Y_i) an individual makes from among the j alternatives depends on his/her latent perception level value Y_i^* relative to thresholds that define the categorical choices from the continuous latent scale γ_j, such that

$$Y_i = j \text{ } if \text{ } \gamma_{j-1} < Y_i^* \leq \gamma_j \tag{2}$$

Therefore, the probability of an individual i selecting alternative j is

$$P(Y_i = j) = \frac{e^{(\beta x_i - \gamma_{j-1})}}{1 + e^{(\beta x_i - \gamma_{j-1})}} - \frac{e^{(\beta x_i - \gamma_j)}}{1 + e^{(\beta x_i - \gamma_j)}} \tag{3}$$

The parameters β_n and thresholds γ_j were estimated using maximum likelihood estimation in SPSS software.

Table 1. Final Estimation Results on RTI Adoption

Parameters	Coeff.	t-stat
Thresholds		
γ_1	−15.444	−3.60
γ_2	−13.357	−3.34
How often do you use paratransit service?		
Once a month (base)		
Few times a month	−9.524	−2.36
Once a week	−19.245	−3.63
Few times a week	−13.216	−3.75
How often do you use paratransit for medical purposes?		
Never (base)		
Rarely	0.446	0.33
Sometimes	−1.703	−0.91
Often	5.528	2.83
How often do you use paratransit for shopping purposes?		
Never (base)		
Rarely	4.945	2.08
Sometimes	−2.897	−0.89
Often	−5.747	−2.78
How often do you use paratransit for work purposes?		
Never (base)		
Rarely	3.208	0.32
Sometimes	5.702	2.11
Often	−4.301	−2.48
How early do you typically schedule your trip?		
One day before (base)		
Less than a week before	1.282	0.91
More than a week before	−0.293	−0.23
What technology do you have access to?		
Computer	8.815	2.50
Smartphone	0.166	0.08
Home phone/ Basic cellphone	−3.301	−1.51
Car	−6.457	−1.65
What is your ethnicity?		
Caucasian (base)		
African American	−9.913	−0.55
What is your income category?		
Less than US$25,000 (base)		
US$25,000 or more	4.634	3.08
What is your age group?		
35–49 years old (base)		
50–65 years old	0.380	0.24
Older than 65 years	−7.418	−2.83

These final parameters can be found in Table 1, and parameters significant at a 95% confidence level are highlighted. A chi-square test showed that the model is significantly better than a constant-only version, at any level of significance.

Overall, the estimation parameters highlight that RTI technology is most likely to be adopted by younger users who have experience using technology and use paratransit for non-routine trips. More importantly, the results confirm that while the current older paratransit riders (who are not familiar with technology) may not be interested in or benefit from RTI technology, future generations of users who *are familiar* with technology would be interested in and benefit from this technology.

Specifically, the "usage" parameters show that the more routinely or frequently individuals use paratransit, the less likely they would be to adopt RTI technology. Additionally, considering different reasons for using paratransit, those who use it often for medical purposes, rarely for shopping and sometimes for work are more likely to adopt RTI technology whereas those who use it often for shopping or work are less likely to adopt the technology. Non-routine tripmaking would benefit most from RTI technology, as there is more uncertainty associated with these trips which could be avoided with the additional information, like vehicle real-time arrival updates, compared with routine trips where there are existing expectations and uncertainty is lower.

Interestingly, usage is not correlated with age (the highest correlation being 2.7%), so these factors are not affected by inexperience with technology either.

While variability of trip purposes affects adoption, the time between the actual trip and when it was scheduled does not. The authors expected this to be important, as one of the main aspects of RTI technology is assistance with scheduling trips. It is possible that scheduling trips with LRPT is easy enough that travelers do not feel this is a concern, relative to other transit providers. Again, scheduled time ahead has only a 10% correlation with age, further supporting the fact that this applies to all users.

Access to and experience with technology was significantly important as well, with access to home computers increasing the likelihood of using RTI, independent of age. Access to smartphone or home phone was prevalent in the surveyed riders but did not have an effect, and access to personal vehicle decreased the likelihood of using the technology. This is to be expected, as those who are comfortable using Internet-connected computers are more interested in technology that could potentially offer a benefit. Without past experience with technology, riders may not be able to appreciate the benefits of such technology, scared of using this technology, or simply not interested in it.

Finally, the demographic variables further emphasize that riders' experience with technology is a key component for adoption. Riders with higher incomes and those aged 65 years or younger are more interested in RTI technology, which is tied to having more discretionary funds to purchase devices and greater confidence using technology. Ethnicity did not affect adoption.

Potential Adoption Score Performance Measure

In order to apply the results of the survey and the ordered logistic regression for transportation planning in small and medium-sized communities, a performance measure for scoring the potential adoption of RTI technology was developed. This performance measure equation utilizes selected weights found in the model for commonly available (and most important from the equation) variables to describe RTI technology adoption in communities and regions. As such, the performance measure scores areas based on three commonly available data characteristics for individuals aged 65 years or older (the age group most tied to paratransit use (*23*, *24*)): computer access, personal vehicle access, and income:

$$\begin{aligned} &\textit{Potential Adoption Score} = \\ &0.443 \times (\%\ \textit{of Population Aged 65 or Older with Computer Access}) + \\ &0.324 \times (\%\ \textit{of Population Aged 65 or Older without Personal Vehicle Access}) + \\ &0.233 \times (\%\ \textit{of Population Aged 65 or Older with Incomes of } \$25k \textit{ to } \$50k) \end{aligned} \tag{4}$$

The weights in the performance measure are derived from the coefficients in the model so they proportionally rank each variable as part of 100% (e.g., 0.443 = 8.815/

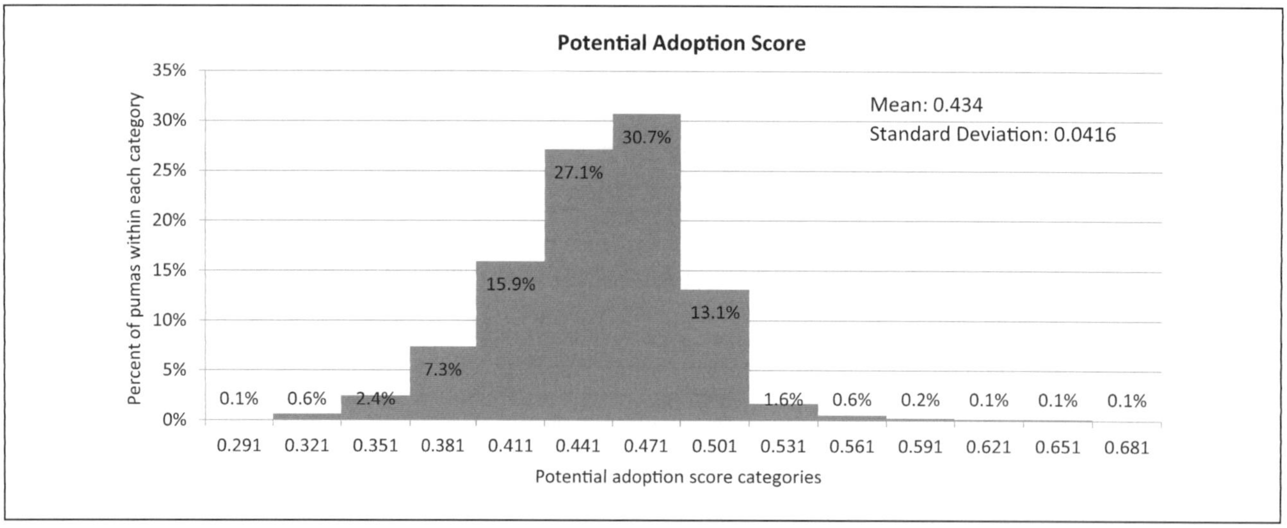

Figure 5. Potential adoption score histogram, based on public use microsimulation areas (PUMAs).

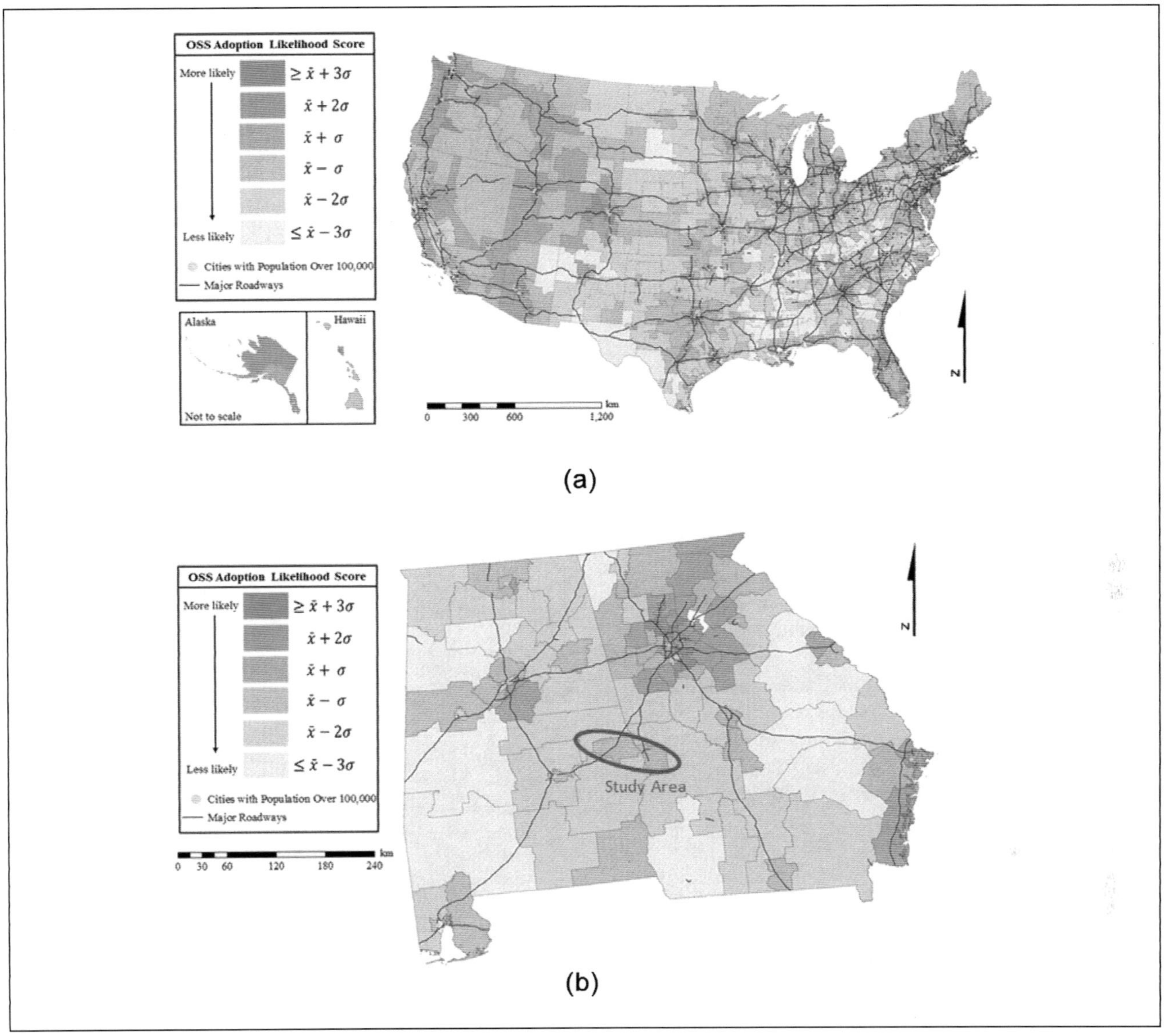

Figure 6. RTI paratransit technology adoption score results: (*a*) country-wide public use microsimulation areas (PUMA) adoption scores; (*b*) Alabama and Georgia PUMA adoption scores.

(8.815 + 6.457 + 4.634). The higher the score for a community, the higher the likelihood of more people adopting the technology. As noted previously, current older populations are less likely to adopt RTI technology, so this performance measure is even more effective in long-term transportation planning and applied to future older populations (i.e., currently 45–55 years old) who are more comfortable using technology.

Data from the 2014 American Community Survey was collected for the 2,351 public use microdata areas (PUMAs) in the United States (*25*). This data was collected and "potential adoption scores" were calculated for each PUMA. Performance measure scores ranged from 0.263 (least likely area for adopting RTI technology) to 0.665 (most likely area for adopting RTI technology) in a nearly normal distribution (seen in Figure 5) with a mean of 0.434 and a standard deviation of 0.0416. Performance measure scores for PUMAs across the country can be seen in Figure 6.

Figure 6 shows that regions in the northeast, the west coast, and Florida show generally higher adoption likelihood scores, whereas many PUMAs in the south and midwest have lower scores. In addition, most metropolitan areas near cities show higher adoption likelihood scores than their surrounding regions. Adoption scores are highly correlated with population size for the PUMA (roughly 70%), so without applying the performance measure calculation, one might generalize these results to focus on implementation in more densely populated regions.

Urban areas show the greatest correlation with potential future growth for paratransit technology. Focusing on Alabama and Georgia, PUMAs in the Auburn-Opelika, Huntsville, Atlanta, and Savannah metropolitan areas show the highest scores. Cities and metropolitan areas have the most potential paratransit growth. These are areas where paratransit use is already concentrated, so this result indicates that the relatively minor cost of investing in RTI technologies can lead to increased ridership in the future.

Conclusions

There are ample opportunities to improve the mobility and access of paratransit users through the use of RTI, and this research seeks to (a) understand the factors affecting who would adopt RTI paratransit technology and (b) provide a planning performance measure that can identify which areas would benefit most from the introduction of such RTI technology. These goals are addressed by modeling paratransit riders' likelihood of adopting RTI technology as well as assessing the benefit communities would see from implementing this technology using on-board passenger survey data collected in two small and medium-sized communities in Alabama and Georgia. Transit agencies will be able to use this information to determine whether RTI would be appropriate for their community as well as the populations that should be targeted for this introduction.

The ordered logistic regression highlighted that access to a computer with Internet and having an annual household income higher than US$25,000 were associated with increased interest in RTI technology on paratransit, while access to a personal vehicle decreased interest. While current older paratransit riders (who are not familiar with technology) may not be interested in or benefit from RTI technology, future generations of users who *are familiar* with technology would be interested in and benefit from this technology. Access to computers and smartphones increases riders' interest in using technology, as do higher incomes. This makes sense, as those more familiar with smart devices would be more likely to use them for transit. Metropolitan areas showed the greatest potential for paratransit growth via RTI technologies.

Additionally, the adoption score performance measure is appropriate for determining communities' potential interest in developing RTI technology for their paratransit ridership prior to undertaking a large study or infrastructure investment. By applying the model results to their local populations (both today and even more appropriately in the future), planners can evaluate where populations are located that are interested in RTI technology for paratransit. Planners can use this information to target these users specifically as transit agencies begin to adopt new technologies.

With such potential benefits of RTI technology, future work should expand on the mechanisms that would assist paratransit riders in using RTI technology and how this technology should function. Additionally, further work should be done with riders from across the country, including looking at participants in areas where RTI technology is already implemented.

Author Contributions

The authors confirm contribution to the paper as follows: study conception and design: J. LaMondia, T. Gajkowski, V. Ramirez; data collection: T. Gajkowski, V. Ramirez; analysis and interpretation of results: J. LaMondia, T. Gajkowski, V. Ramirez ; draft manuscript preparation: J. LaMondia, T. Gajkowski, V. Ramirez. All authors reviewed the results and approved the final version of the manuscript.

References

1. Fei, D. *The Americans with Disabilities Act of 1990 (ADA) Paratransit Cost Issues and Solutions for the Greater Richmond Transit Company (GRTC) and Beyond.* Master of Science thesis. Virginia Commonwealth University, Richmond, VA, 2014.
2. Federal Transit Administration. *Enhanced Mobility of Senior & Individuals with Disabilities - Section 5310.* U.S. Department of Transportation Washington, D.C., https://www.transit.dot.gov/funding/grants/enhanced-mobility-seniors-individuals-disabilities-section-5310. Accessed July 17, 2016.
3. Koffman, D., R. Weiner, A. Pfeiffer, and S. Chapman. *Funding the Public Transportation Needs of an Aging Population.* Nelson/Nygaard Consulting Associates, San Francisco, CA, 2010.
4. Ortman, J. M., V. A. Velkoff, and H. Hogan. *An Aging Nation: The Older Population in the United States.* Publication P25-1140. United States Census Bureau, Suitland, MD, 2014.
5. Goodwill, J. A., and H. Carapella. *Creative Ways To Manage Paratransit Costs.* Center for Urban Transportation Research, University of South Florida, Tampa, FL, 2008.
6. Zhang, F., Q. Shen, and K. Clifton. Examination of Traveler Responses to Real-Time Information About Bus Arrivals Using Panel Data. *Transportation Research Record: Journal of the Transportation Research Board,* 2014. 2082: 107–115.
7. Brakewood, C., G. S. Macfarlane, and K. Watkins. The Impact of Real-Time Information on Bus Ridership in New York City. *Transportation Research Part C: Emerging Technologies,* Vol. 53, 2015, pp. 59–75.

8. Tang, L., and P. V. Thakuriah. Ridership Effects of Real-Time Bus Information System: A Case Study in the City of Chicago. *Transportation Research Part C: Emerging Technologies*, Vol. 22, 2012, pp. 146–161.
9. Greater Cleveland Regional Transit Authority. *GCRTA Online ADA Services*, Cleveland, OH, 2016, http://paratransit.gcrta.org/Accessed July 28, 2016.
10. *TCRP Synthesis 91: Use and Deployment of Mobile Device Technology for Real-Time Transit Information*. Transportation Research Board of the National Academies, Washington, D.C., 2011.
11. Politis, L., P. Papaioannou, S. Basbas, and N. Dimitriadis. Evaluation of a Bus Passenger Information System from the Users' Point of View in the City of Thessaloniki, Greece. *Research in Transportation Economics*, Vol. 29, 2010, p. 250.
12. Borole, N., D. Rout, N. Goel, P. Vedagiri, and T. V. Mathew. Multimodal Public Transit Trip Planner with Real-Time Transit Data. *Procedia: Social Behavioral Sciences*, Vol. 104, 2013, pp. 775–784.
13. Cats, O., and G. Luotos. Evaluating the Added-value of Online Bus Arrival Prediction Schemes. *Transportation Research Part A*, Vol. 86, 2015, pp. 35–55.
14. Watkins, K. E., B. Ferris, A. Borning, G. S. Rutherford, and D. Layton. Where Is My Bus? Impact of Mobile Real-time Information on the Perceived and Actual Wait Time of Transit Riders. *Transportation Research Part A*, Vol. 45, No. 8, 2011, pp. 839–848.
15. Brakewood, C., S. Barbeu, and K. Watkins. An Experiment Evaluating the Impacts of Real-Time Transit Information on Bus Riders in Tampa, Florida. *Transportation Research Part A*, Vol. 69, 2014, pp. 409–422.
16. Bregman, S., K. Watkins, Y. A. Xu, and K. Coffel. Who's Online...and Who Isn't? Equity Considerations for Web-Based Feedback and Transit. Presented at 94th Annual Meeting of the Transportation Research Board, Washington, D.C., 2015.
17. Windmiller, S., T. Hennessy, and K. E. Watkins. Accessibility of Communication Technology and the Rider Experience. *Transportation Research Record: Journal of the Transportation Research Board*, 2014. 2415: 118–126.
18. Zickhur, K. *Who's Not Online and Why?* Pew Research Center, Washington, D.C., 2013.
19. Finn, B. Towards Large-scale Flexible Transport Services: A Practical Perspective from the Domain of Paratransit. *Research in Transportation Business & Management*, Vol. 3, 2012, pp. 39–49.
20. Davison, L., M. Enoch, T. Ryley, M. Quddus, and C. Wang. Identifying Potential Market Niches for Demand Responsive Transport. *Research in Transportation Business & Management*, Vol. 3, 2012, pp. 50–61.
21. Fei, D., and X. Cheng. The Americans with Disabilities Act of 1990 (ADA) Paratransit Cost Issues and Solutions: Case of Greater Richmond Transit Company (GRTC). *Case Studies on Transport Policy*, Vol. 3, No. 4, 2015, pp. 402–414.
22. Lee-Russell Public Transit. *Lee-Russell Public Transit Passenger Guide*, Auburn Opelika Metropolitan Planning Organization, Opelika, AL, 2010.
23. American Public Transportation Association. *A Profile of Public Transportation Passenger Demographics and Travel Characteristics Reported in On-Board Surveys*. Washington, D.C., 2007.
24. Multisystems, Inc. *Strategies for Improved Traveler Information*. TCRP Report 92. Multisystems, Inc., Cambridge, MA, 2003.
25. U.S. Census Bureau. *2010 Public Use Microdata (PUMAs) and Characteristics of Microdata*. U.S. Census Bureau, Washington, D.C., 2013.

The Standing Committee on Transportation Planning for Small and Medium-Sized Communities (ADA30) peer-reviewed this paper (18-00831).

Article

TRR
JOURNAL OF THE TRANSPORTATION RESEARCH BOARD

Transportation Research Record
2018, Vol. 2672(51) 68–78

DOI: 10.1177/0361198118787639
journals.sagepub.com/home/trr
SAGE

Strategically Aligning Capital Improvement Prioritization to Performance Goals

Elliot Sperling[1] and Catherine Ross[1]

Abstract

In complying with federal reporting requirements, state transportation agencies will need to demonstrate progress toward achieveing national performance targets in their next long-range transportation plan (LRTP) and state transportation improvement program (STIP). In order to better understand current practices a survey was administered to state departments of transportation (DOTs) to gauge their overall status in complying with rulemakings on transportation performance management, and to measure DOT interest in using new metrics and prioritization techniques to advance data-driven processes for decision making. Furthermore, it was to examine evidence of cross-modal project evaluation and to explore practices by which states align decision criteria and analysis techniques with national performance goals and objectives. Survey results revealed the transition in practice toward performance-based approaches, but more limited use of cross-modal evaluation criteria. Case studies were also used to document decision criteria areas that help states advance alignment of their planning and programming processes. From the survey and case study analysis, recommendations are provided to underscore the need for multi-criteria decision analysis, or a similar decision analysis framework, to help states align project evaluation with national performance goals. As state DOTs adopt performance-based approaches they must review their project selection processes to ensure performance measures are linked to current investments, and funded projects align with agency goal areas and desired outcomes.

State transportation agencies are responsible for investing public resources in funding, managing, and operating many of the nation's significant transportation assets. Although state departments of transportation (DOTs) address fundamental priorities regarding safety, mobility, maintenance, and operation of their respective transportation systems, they also have to consider many indirect goals: supporting economic development, improving quality of life, providing connections to vital services and employment, and mitigating environmental impacts, among others (*1–4*). At the same time, agencies face funding restrictions, programmatic limitations, public desires, and political realities. Within this context, performance-based approaches offer opportunities to address these challenges simultaneously.

The current state of practice largely reflects silo-based methods, particularly in programming, in which resource allocation decisions often fail to account for system-wide or agency-wide implications. To a large extent, the focus has been on program areas or funding by geographic districts, rather than more comprehensive approaches for project evaluation (*5*).

In order to validate or better understand the current state of practice across the nation, we implemented a survey of state DOTs that led to the development of four case studies. The survey focused on three main challenges in programming:

1. Addressing transportation needs across geographies;
2. Prioritizing projects across programs and divisions;
3. Funding across travel modes.

The survey was conducted at a time when the national focus began shifting toward performance-based management, which centers on outcomes as opposed to processes. This shift was clearly reflected in survey results. For both capital programming and planning, states express growing interest in directly aligning these processes with national performance goals. Given that state DOTs differ in their structure of divisions, use of funding programs, and process requirements, state evaluation approaches were broadly examined through the survey.

[1]Georgia Institute of Technology, School of City and Regional Planning, School of Civil and Environmental Engineering, Atlanta, GA

Corresponding Author:
Address correspondence to Elliot Sperling: elliot.a.sperling@gmail.com

Four state DOT case studies are presented to show evidence of multi-criteria evaluation in the following areas: 1) a statewide funding program, 2) a qualitative assessment framework for capital enhancements, 3) a quantitative assessment framework for capital enhancements, and 4) a mixed analysis assessment framework in evaluating projects across modal and federal program areas. Taken together, these four approaches guide recommendations using multi-criteria evaluation approaches for project assessment based on national performance goal areas.

Federal Legislative Background

Until the MAP-21 legislative reauthorization in 2012, state DOTs did not need to demonstrate performance progress to the Federal Highway Administration. Each state transportation agency developed its own approaches to manage their systems, thus leading to large variation in how states manage their assets. The degree to which their long-range plans influenced their short-term transportation investments was entirely up to the discretion of each state (*6*). MAP-21 requires that state DOTs incorporate performance management principles into a number of formal plans and planning processes, including: long-range transportation plans (LRTPs) and state transportation improvement programs (STIPs). The final rulemaking on statewide, metro planning, and non-metro planning requires a linkage between planning and programming. States are to: establish performance targets, then include those performance targets in plans, which they then use to link their investment priorities to targets, and finally report on progress (*7*).

Under the Fixing America's Transportation Act (FAST Act) there is an emphasis on freight, with two new national freight funding programs providing close to US$10 billion in funding over 5 years. Both the INFRA grant program and the National Highway Freight Program (NHFP) have an intermodal component, but funding for non-highway freight improvements is restricted and cannot exceed more than 10 percent of total funds (*8–10*). The inclusion of freight programs is a notable change from prior surface transportation authorizations, in which previous appropriations offered no dedicated sources of funding for freight improvements. As a result, states are producing formal plans that identify freight needs. Survey results indicated that most states have a channel whereby freight interests identify their priorities, which are then connected to project-evaluation decisions.

Literature Review of Performance-Based Emphases to Mobility Investments

Although performance-based emphases are largely discussed in a post-MAP-21 environment, states have a longstanding history of using performance measurement criteria in the areas of safety and asset condition as well as in measuring congestion and delay (*2, 11*). The U.S. Government Accountability Office surveyed all states in 2010 prior to the implementation of MAP-21. Survey results indicated challenges such as identifying indicators for livability and collecting data across transportation modes that limited a broader use of performance measures. Despite these challenges, several state DOTs evidence the testing of new approaches that connect evaluation methods with performance-based goal areas (*12*).

Strengthening Planning and Programming Linkages

Impediments continue to inhibit more coordinated agency linkages between what are two distinct processes: planning and programming. Update cycles for plans and programs can differ substantially, with plans updated over several years, as compared to the use of regular budget cycles for programming (*13*). In addition, organizational responsibilities are typically separate, with programming functions carried out by different divisions within an agency other than planning, and with a lack of interface between them (*4*). Agencies have also used evaluation criteria that promote little or no internal consistency between the multimodal planning desires espoused, and the programming decisions made. With modal decisions decided independently in the planning process, states may encounter difficulty in implementing cross-investment approaches, where an agency can easily move funds between highway, transit, and intermodal project types to address needs. Examining modal tradeoffs and promoting travel options has lacked clear examination in most planning and programming approaches used today (*5, 14*).

Overcoming Modal and Programmatic Barriers

Many state DOTs continue to use programmatic categories to prioritize projects for funding, where mode-specific "silo" approaches are commonplace in the development of most states' STIP (*2, 3*). This situation is largely due to challenges faced in comparing projects across modes using traditional metrics that are widely-used and validated (*2, 3*). The existence of programmatic barriers and set funding formulas, that is, the longstanding use of programmatically-restrictive block grant apportionments by the Federal Highway and Federal Transit Administrations, can prevent more optimal distributions of funds to address the travel needs of all transportation system users (*15–17*).

The present use of "stove-piped" funding sources and organizational separation of modal divisions limits the ability to evaluate the tradeoffs of funding decisions, and to redirect levels of investment across different travel modes (*5, 14*). Many plans espouse or follow a multimodal philosophy in policy orientation, but the achievement of multimodal goals tends to lag behind the achievement of other goal areas

(*14*). Many state agencies deemphasize a multimodal programming link because of modal separation in decision making and the prevalence of funding restrictions that hamper efforts toward multimodal decision-based approaches (*5*).

The National Cooperative Highway Research Program Report 806 distinguishes state practices based on four stages. **Legacy driven** states approach resource allocation through existing program structures. Funding changes are generally at the margins, adjusted for inflation or changed priorities. **Fix it first** states allocate funds based on current asset management systems and their calculated preservation needs. Financial constraints leave less funding for capital improvements, and so remaining allocations tend to reflect desired projects, versus projects that best align with strategic goal areas. **Soft optimization** states have flexible allocations, adjusted to be better reflective of relative priorities, but such decisions are based on professional judgment, DOT leadership, or non-technical inputs. **Performance-based** states have practices that allow for cross-asset allocation decisions to be made, as reflected through performance measurement tools. Decisions are mostly data-driven through use of models and tools that forecast outcome achievement that aligns with agency goals. Allocation decisions are still based on professional judgments in balancing tradeoffs appropriately (*5*).

Although states find themselves at different stages, a common remedy to the programmatic challenges faced is through use of multi-objective criteria for decision analysis (*18*). This remedy is one that was recommended based on the survey findings, review of state practices, and current literature (*5*). Though other methods exist, many states express difficulty in using alternative tools consistently across project types. For instance, benefit-cost analysis (BCA) is difficult to use as a standalone tool; it does not allow agencies to assess the degree to which different projects connect with agency objectives and goal achievement.

Decision Science Applications and Trade-Off Analysis

Beyond congestion, delay, safety, and asset condition metrics, there are other impacting factors for capital improvements, regarding how efficiently such projects generate returns in economic development, promote access to vital destinations, provide connections, and mitigate impacts that affect environmental and quality of life factors (*4*). Project impacts cannot all be readily quantified in monetary terms; however, these transportation impacts are largely reflected in policy-based goals and objectives (*1*), and can more closely connect to decision making. Using both quantitative and qualitative measures in multi-criteria decision analysis techniques allows for impacts to be understood in a more objective manner (*18*).

Several factors may be considered in determining whether objective-based approaches will make the process of decision making more effective, including: the degree that the decision-making approach instills process accountability on the use of agency resources, provides for flexibility (in removing funding barriers) to allow for wider sets of tradeoffs and programmatic allocations to be considered, and also considers multijurisdictional and multimodal coordination and planning to fund projects addressing challenges at various scales—local, regional, and statewide (*4*, *19*). Additionally, many agencies would like to see cost-effectiveness receive heightened consideration in the process.

National Survey Results on State DOT Practices

In early 2017 an online survey was sent to targeted points-of-contact in all 50 state transportation agencies with senior level roles in planning, programming, or policy. The survey was available through an online link for six-weeks, and during that time 35 DOT respondents completed the survey (*20*), as shown in Figure 1. Respondent titles are largely encapsulated under the positions of Administrator, Assistant Director of Planning, Bureau Chief of Programming, Director of Project and Planning Development, and Statewide Planning Engineer.

State respondents noted difficulties in addressing three challenges referenced in the survey: the geographic distribution of funds, comparing project alternatives across funding programs, and mode-neutral project evaluation (use of person-based and ton-based metrics to compare across non-preservation projects).

Of the challenges referenced, states encounter significantly more difficulty in developing mode-neutral emphases. Funding presents a large barrier, and based on the commonality of responses, federal programmatic restrictions are a barrier in agencies confronting a lack of funds, or flexibility, to address asset needs from the national highway system (NHS).

To understand modal challenges associated with funding limitations, a survey question listed seven different project types. Respondents were asked to rank the ones most difficult to fund based on existing funding levels and sources of funding available. Respondents ranked these project types from 1 to 7, with 7 being the most challenging to fund. For simplicity, overall findings are normalized: 1 indicates the most challenging project type to fund and 0 indicates unanimous consensus that a project category is easiest to fund. This ranking procedure led to the following normalized scores:

- Intermodal freight: rail/port/transfer facilities – 1.00
- Multimodal transfer facilities/connections – 0.91
- Travel demand management (TDM) – 0.74

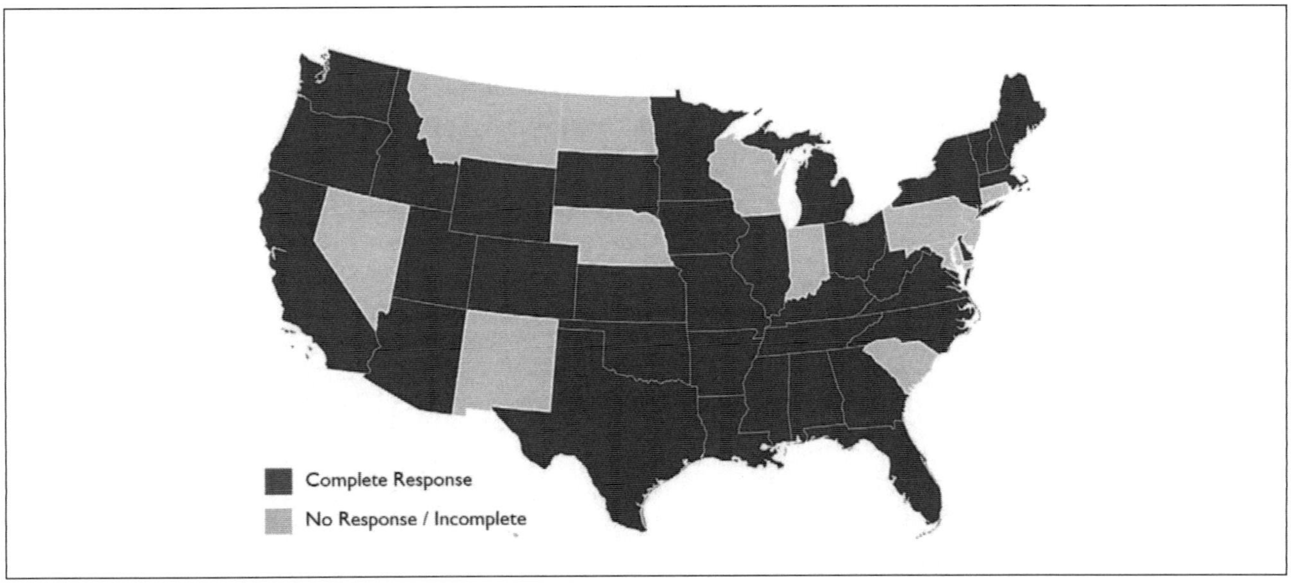

Figure 1. Survey participation (35 state DOTs).

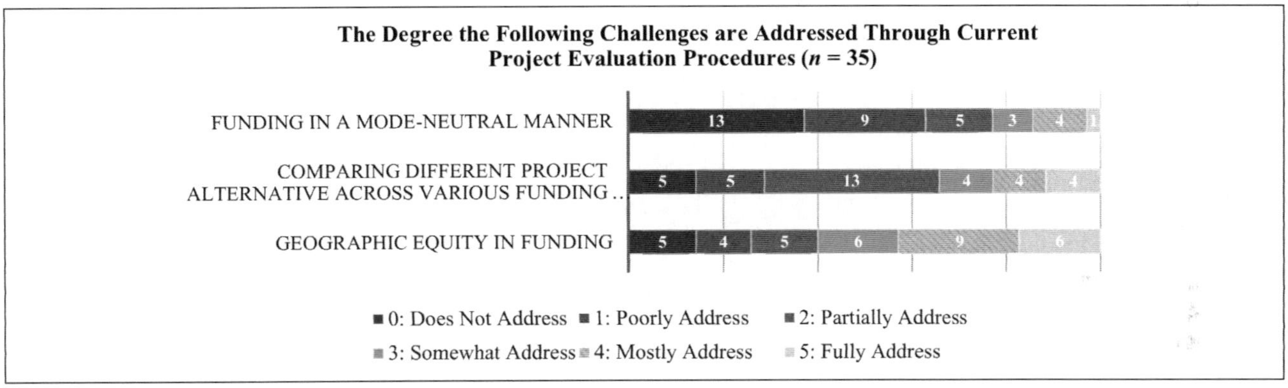

Figure 2. Challenges in distributing funds by area, program, and mode in project-evaluation procedures.

- Transit – 0.68
- Bike and pedestrian facilities – 0.68
- Roads off the national highway system (NHS) – 0.46
- Highways – 0.06

Projects involving multimodal and intermodal interactions were reported as the most difficult to undertake. Based on open-ended responses and survey questions, adequate funding and evaluation tools that are applicable to them appear to be deficient. One agency believes it commits to projects in a fully objective manner across modal asset categories, whereas 22 of the 35 state DOT respondents indicate that current project-evaluation procedures either do not address or poorly address funding in a mode-neutral manner, as shown in Figure 2 below (*20*).

When agency respondents were asked about a specific way to improve funding flexibility through the federal program, most were supportive of increased flexibility for both the NHFP and INFRA Grant Program. Fifteen states (42.9%) selected "unsure," indictating they may want to understand the effects of such a proposal more fully. Four (11.4%) were not supportive of removing the 10% restriction, and one preferred not to answer. By a 3.75:1 margin, states taking a position are in support of having more non-highway specific funds for freight projects to address intermodal needs in their state.

On states' overall project-evaluation approaches, 19 states indicate that they fund capital improvement projects across different funding programs at the federal level. A survey question followed that asked about some of the factors considered for capital improvement prioritization (*20*). This question listed factor areas generally to encapsulate criteria more readily used to evaluate capital improvement projects, for which federal funding is being pieced together to support

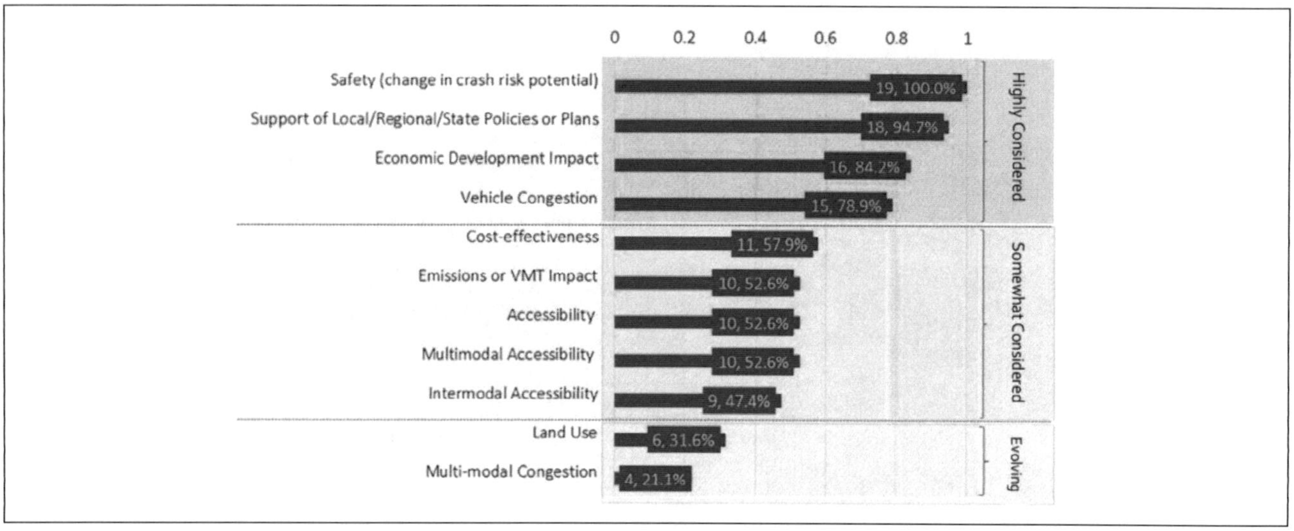

Figure 3. Factors considered in project evaluation across programs.

Table 1. Crosstabs Analysis: Changes in Interest in Adopting New Metrics and Whether BCA or CEA is Used in Project Evaluation Prior to STIP Development

Livability, environmental, economic, and social factors	BCA/CEA methods used (n = 24)	Not used (n = 11)
Economic development	2.58	2.64
Disaggregated crash data by travel mode	2.26	0.73
Intermodal connections	1.88	1.36
Multimodal accessibility	1.88	1.09
Health / active transportation	2.00	0.45
Equity (disadvantaged populations)	1.79	0.73
Air quality	1.13	0.73
Statewide VMT impact	0.74	1.55
Land use coordination	0.79	0.18
Non-mode specific congestion or mobility measures (person-mile and ton-mile movement)	0.87	-1.18
GHG emissions impact	0.61	-0.18

Note: Rating average computation: not interested = −5/less interest = −3/no change = 0; more interested = +3/currently used = +5.

these projects. For this question, states largely considered factors in a qualitative and quantitative sense for how they generalizably evaluate capital improvement projects. As shown in Figure 3 below, there are three clusters evident. Factors more prevalent in project-evaluation decisions include: safety, support of plans, economic development impact, and vehicle congestion-based metrics. Those less prevalent in analysis include: cost-effectiveness, emissions impact or travel change, and accessibility change. Evolving factors of analysis are in considering land use and multimodal congestion-based criteria in project evaluation (*20*).

States that use quantitative-based approaches, cost- effectiveness analysis (CEA) or benefit–cost analysis (BCA) in project evaluation have greater interest in livability, environmental, and social factors in multi-criteria project-level assessment than those that lack monetization evaluation techniques at the project-level, as shown by the summarized rating values in Table 1. States that have no interest in a mentioned area in Table 1 are counted with a -5 score, whereas those currently using metrics in that factor area for evaluating projects receive a +5 score. States more interested in using evaluation criteria in a factor area receive a +3 score, and those less interested receive a -3 score. No change of interest counts as a score of 0. A cross tabs analysis was used to analyze responses to this question and to separate states into two categories: those that use BCA, CEA, or both BCA and CEA, and those that do not. Overall, states with BCA or CEA evaluation approaches in place have greater interest in accounting for project-level impacts on environmental, social, and livability factors, and person-based congestion criteria (required in statewide reporting under the system performance rule, but not under a project basis).

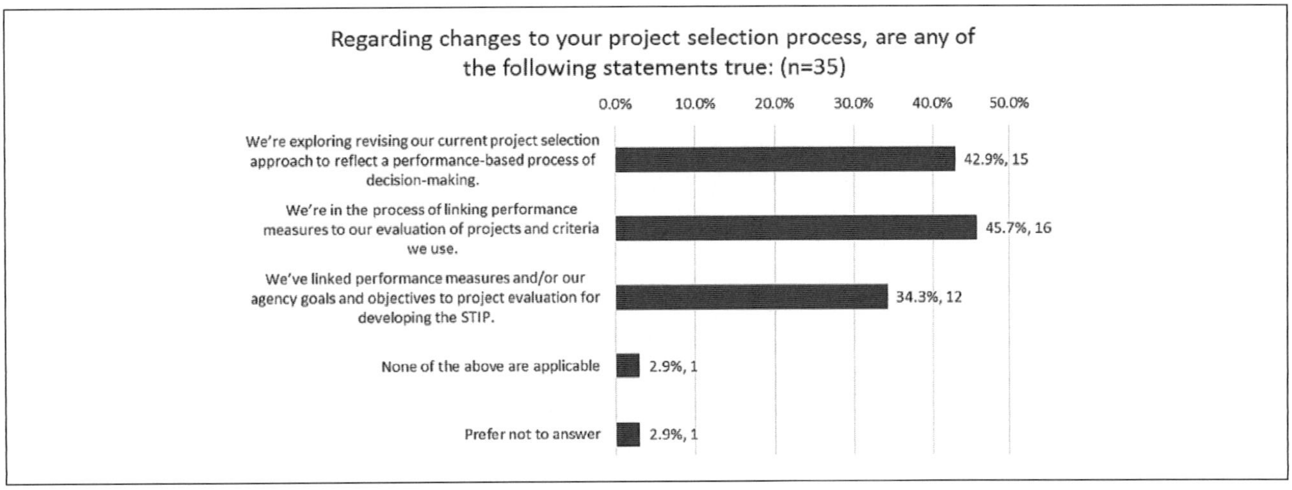

Figure 4. Use and exploration: performance-based and project-evaluation-based linkages.

The extent of factor usage in Table 1 was used as a means to consider multi-objective-based approaches for state DOTs, and to reflect on how a larger use of measurement tools to guide decision making can promote alignment with the seven national transportation performance goals (*20*). According to the survey findings in March of 2017, 22 of the 35 states (62.9%) specified that the rulemakings did not have impact at this point on selection of projects for the STIP. Six states (17.1%) specified it had an effect, whereas seven states (20.0%) provided an "other" response. Of the "other" responses, one specified their next year's STIP will have impact on selection of projects, two mentioned it had some impact, another two mentioned they were prioritizing with their own performance-based approach before MAP-21 rulemaking implementation. One agency made note of the temporary hold on certain rulemakings, implying it may have delayed their implementation of performance-based requirements (*20*).

States are contemplating changes in their project-evaluation processes to address MAP-21 rulemakings for their STIP and LRTP. As noted in Figure 4 below, states are mostly in the position of exploring the revision of their current project selection process, or linking performance measures to the evaluation of projects and criteria used (15 and 16 state DOT responses, respectively, were recorded as being a true statement). For this question, several states indicated that more than one statement was true.

Case Studies

Based on survey responses, four state DOT practices were selected to demonstrate use of multi-criteria project evaluation. Oregon DOT (ODOT) indicated use of qualitative criteria to inform project evaluation through the Enhance Program (*21*). The Enhance Program funds new capital expansions, and though it has a statewide orientation, each ODOT District separately reviews projects through qualitative considerations to move toward a prioritized list of projects. Three modal attribute criteria are considered: 1) connectivity and system benefits, 2) safety and public health, and 3) accessibility and mobility. ODOT also includes cross-modal criteria, with a similar qualitative emphasis focused around economic development, social benefits, environmental stewardship, safety, project readiness, and leverage.

Maine DOT highlighted a specific program they use, focused on freight rail connections through their Intermodal Rail Access Program. The program uses a mix of qualitative and quantitative criteria, requiring projects to show a benefit–cost ratio through a consistently applied methodology, with the overall purpose to ensure public benefits are realized through projects being able to demonstrate reduced highway maintenance expenditures and lessened congestion costs (*22*).

Utah DOT (UDOT) uses a data-driven support system, focused on quantitative factors to assess road projects by type, and to support capital investment prioritization based on objective criteria (*23*). Utah's prioritization process is focused on safety, congestion reduction, size and magnitude of the project, and other benefits, with cost–benefit analysis required for certain project categories, as shown in Table 2. A significant objective of this research was centered on cross-investment prioritiziation approaches using consistent criteria, and whether states have successfully implemented such practices. An example of an evaluation process capable of evaluating modal investments would help to showcase a practice that the literature review found difficult to implement in practice. Given the challenges pointed to earlier in this paper, a lack of multimodal evaluation practices were found through the survey. UDOT, for instance, is well ahead of other states in using quantitative criteria and rankings for

Table 2. Virginia DOT (VDOT) SMART SCALE and UDOT Prioritization Criteria: Capital Projects (*22, 23*)

VDOT SMART SCALE			UDOT capital improvement prioritization (road project-specific)						
Project types: All capital improvements, outside asset management (VA code: §33.2-369)			Project types: All roadway capital improvements that expand capacity (evaluation based on below project types)						
Factor areas	SMART SCALE evaluation criteria	Quantitative/ qualitative evaluation	Quantitative metrics	Widen existing facility	New facility	Upgrade existing at-grade intersection	New interchange on existing freeway	Upgrade existing interchange	Passing lane
Safety	S.1 EPDO of fatal and injury crashes[a]	Quantitative	Safety	Safety index[f] (25%)		Safety index[f] (25%)		Safety index[f] (25%)	Safety index[f] (30%)
	S.2 EPDO rate of fatal and injury crahses[a,b]	Quantitative							
Congestion mitigation	C.1 Increase in person throughput[b]	Quantitative	Congestion scores	Existing V/C ratio (25%)	Parallel facility V/C ratio (30%)	Daily vehicle hours saved (30%)	Daily vehicle hours saved (30%)	Daily vehicle hours saved (30%)	
	C.2 Decrease in person hours delay[b]	Quantitative			Parallel facility V/C improvement (30%)		Adjacent interchange V/C ratio (10%)		
Economic development	ED.1 Consistency with ED plans[d]	Mainly qualitative	Size/ magnitude scores	Roadway volume (20%)	Future roadway volume (25%)	Intersection volume (20%)	Future ramp volumes (20%)	Future ramp volumes (20%)	Roadway volume (30%)
	ED.2 Freight tons/intermodal access	Both							
	ED.3 Travel time reliability	Both		Truck volume (10%)	Future truck volume (15%)		Distance to adjacent intersection (5%)		Truck volume (20%)
Accessibility	A.1 Increase access to jobs[e]	Quantitative							
	A.2 Access to jobs[e] (disadvantaged popls.)	Quantitative		Functional clasification (5%)					
	A.3 # of non-SOV users	Mainly quantitative							
Environment	E.1 Potential to improve air quality and reduce GHGs	Mainly qualitative	Other Scores	Transportation Growth (15%)		B/C Ratio (25%)	B/C Ratio (35%)	B/C Ratio (25%)	Freight Corridor[g] (20%)
	E.2 Minimizes environmental impacts (NEPA process)	Mainly Qualitative							
Land use	L.1 Transportation-efficient land use	Mainly qualitative							
Cost-effectivess used to weight final score; Regions weight factor areas			Weights are indicated above, and calculated into normalized scores out of 100 points						

[a]Crash modification factors (CMFs) are used to estimate equivalent property damage only (EPDO).
[b]Rate by reduction of crashes per 100 million VMT.
[c]Volume-to-capacity (V/C) ratio of 0.95 has to be achieved in the no-build scenario for a score to be calculated.
[d]Access to jobs based on cumulative accessibility within 45 min (driving), 60 min (transit).
[e]Economic development (ED) plans and policies.
[f]Safety index (SI) = crash rate score + 3 (severity score) – 2 (SI range = 1 to 10).
[g]Freight corridor score based on roadway classification.

making data-driven investment decisions on large capital projects that are roadway and highway-based, but, like the vast majority of state DOTs, they face challenges in developing a mode-neutral emphasis. They use more of a soft optimization approach to address modal alternatives outside their prioritization process. Practices emphasizing mode-neutral decision-making remain rare, despite federal performance requirements moving in this direction. Most of the states that commented specifically about mode-neutral project evaluation mentioned it was good in theory, but simply too hard for them to implement.

Recently, VDOT developed a mode-neutral capital project prioritization process that they refer to as SMART SCALE. Based on survey responses and a critical review of approaches mentioned in responses, Virginia was alone in emphasizing a mode-neutral approach to project evaluation across capital project funding in using a consistent defined set of criteria to evaluate highway, rail, transit, TDM, bike, and pedestrian facility additions. This led to their criteria being emphasized more substantially in the recommendations. The project-evaluation metrics the state uses emphasize person-based movement measures for congestion mitigation. Accessibility is tied to both highway and transit investments. Furthermore, several of the other categories take into account the provision of multimodal travel options in scoring criteria. The approach focuses on outcome-based and objective analysis of capital projects, and instills a goal achievement framework into project evaluation (*24*). Six factor areas are used, shown in Table 2, with the exclusion of land use consideration in rural regions of the state. SMART SCALE emphasizes a collaborative process of regions and local entities responsible for submitting project applications. Virginia's Commonwealth Transportation Board uses the same technical evaluation criteria for all qualified projects seeking state funding. To minimize the risk of subjectivity and political influence, a QAQC (quality analysis and quality control) review takes place to ensure fairness and objectivity in reporting final scores. All factor areas are scaled by the regions, to ensure projects are evaluated by the benefits each project provides to an area. The anticipated cost of a project is factored into the final benefit score to ensure that cost-effectiveness is a critical factor in deciding what state investments to make.

Taken together all four of these state DOT practices provide models for how project evaluation can move toward more of a performance-based linkage, as states look to address challenges in programming across geographies, federal programs, and travel modes in a more transparent, data-driven account of project benefits to ensure that capital enhancements have long-term benefit. Capital improvements are particularly suited toward multi-criteria decision analysis evaluation in assessing whether long-term investment decisions align with performance goal areas. Given capacity expansion projects are additive, they can pose higher long-term costs for DOTs to achieve a state of good repair. Thus, the overall value of each of these improvements versus the request for funds is important to assess.

State and National Policy Recommendations

Survey results suggest states are in an interim period. MAP-21's national performance goals were only recently translated into final performance rulemakings in early 2017. The survey finds that the vast majority of states are in the process of aligning their planning and programming approaches to ensure that future plans and programs can meet or exceed the objectives of the national transportation performance final rules. During this interim period, agencies are looking at ways to demonstrate how their state investments translate into achieving performance targets. Linkages in planning and programming remain poorly defined in many state decision-making processes, and without clear linkages with performance-based emphases, use of traditional approaches to project evaluation may reduce the effectiveness that performance measurement has for agencies. Many states continue to keep separate planning and programming divisions; without high-level direction in agency departments to ensure coordination between plans and programs, this approach can lead to disjointed outcomes.

Evaluative criteria for selecting capital improvement projects can better connect with national performance goals and objectives, and the state's LRTP.

To analyze project-level impacts more fully across all national performance goal areas referred to in MAP-21 legislation, overall criteria areas are suggested for states to consider in linking their project-evaluation process to six of the seven national performance goals from MAP-21 (*reduced project delivery delay* relates to the later procurement stage of a project, and thus is not included in the criteria below). An additional performance goal area is recommended: *system accessibility*, a result emanating from the survey findings of states demonstrating interest in accessibility benefits being evaluated to help guide capital improvement project prioritization. The specific evaluative criteria measures originate from an in-depth examination of specific programs and overall practices from the four states examined in this research: Maine, Oregon, Utah, and Virginia. The criteria areas directly link to national goals, as referenced in Figure 5.

- **I. Safety**: $S\text{-}_{EV.1}$ Reduction in Fatal and Severe Crash Potential
- **II. Infrastructure Condition**: $I\text{-}_{EV.1}$ Pavement/Bridge Condition; $I\text{-}_{EV.2}$ Lifespan Reached; $I\text{-}_{EV.3}$ Criticality (annual average daily traffic, system importance measure, or index)
- **III. Congestion Reduction**: $C\text{-}_{EV1}$ Decrease in Total Person Hours of Delay
- **IV. System Reliability**: $R\text{-}_{EV.1}$ Increase in Person Throughput;
  $R\text{-}_{EV.2}$ Travel Time Reliability Improvement

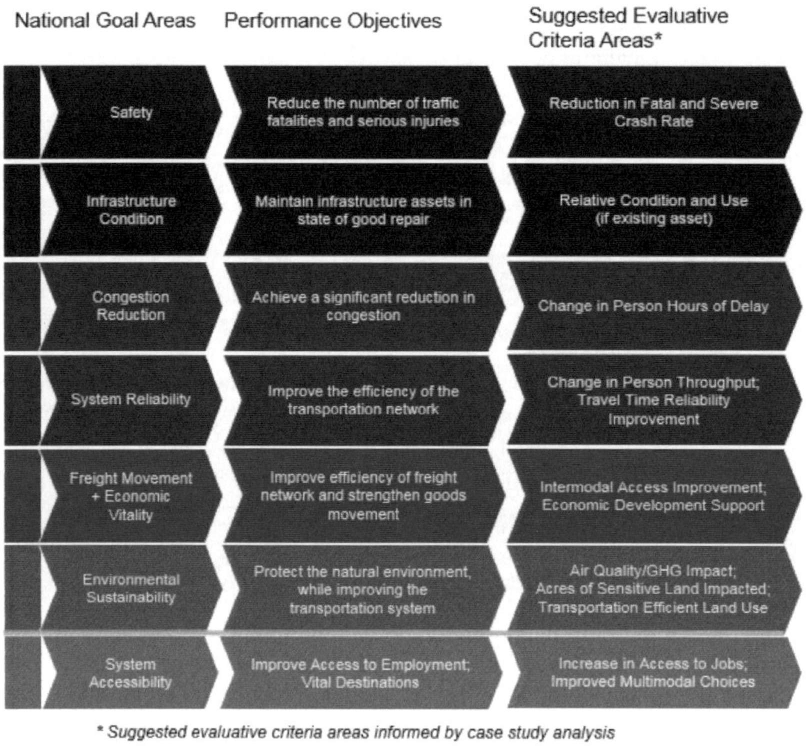

Figure 5. Linking national performance goal areas to state-based project evaluation.

V. **System Accessibility**: A-$_{EV.1}$ Increased Access to Jobs/Critical Services;
A-$_{EV.2}$ Improved Access to Multimodal Choices

VI. **Freight Movement and Economic Vitality**: F-$_{EV.1}$ Ton-Mile Movement;
F-$_{EV.2}$ Intermodal Access; F-$_{EV.3}$ Economic Development Support

VII. **Environmental Sustainability**: E-$_{EV.1}$ Air Quality/GHG Emissions Impact; E-$_{EV.2}$ NEPA Process Anticipated * Acres of Environmentally-Sensitive Land Impacted;
E-$_{EV.3}$ Transportation-Efficient Land Use

Suggested evaluation criteria measure areas S-$_{EV}$, C-$_{EV}$, R-$_{EV}$, A-$_{EV}$, and E-$_{EV}$ are based on the SMART SCALE process used by the Commonwealth of Virginia. Portions of evaluation criteria measures I-$_{EV}$ and F-$_{EV.1, 3}$ are recommended based on UDOT's capital improvement approaches and other asset management emphases. Maine's Intermodal Rail Access Program informed the F-$_{EV}$ evaluative criteria area (*25*). Additionally, portions of measures S-$_{EV}$, A-$_{EV}$, and E-$_{EV}$ are emphasized based on ODOT's *Enhance* qualitative criteria (*21*, *22*, *23*, *24*).

The evaluative criteria mentioned above can be considered under a multi-criteria decision analysis framework. To consider them, states will need to examine their existing data sources and other available data to tie project-evaluation decisions to more detailed evaluation tools, and to obtain consensus on weighting these different objectives to compute final scores.

Findings from the survey suggest there is opportunity to better align evaluation criteria with the outcome-based intents of performance goals stemming from MAP-21 legislation. States continue to lack multimodal-based metrics for evaluating congestion reduction and system reliability goals, pertaining to cross-asset project-evaluation decisions. Many states are also less experienced with the linkages of freight planning and programming. For environment and sustainability factors many states may qualitatively understand impacts, but it is rare for states to use evaluation measures consistently—particularly outside of the Congestion Mitigation and Air Quality Improvement (CMAQ) program, in which, specific to that program, states in non-attainment are required to demonstrate air quality benefits. States also present varying opinions and approaches in addressing environmental criteria; the survey found large differences in agency perspectives, particularly over whether to assess greenhouse gas emissions' impact in project evaluation. States through their agency respondent, however, expressed more interest than disinterest in addressing the social, livability, and environmental factors asked about, and whether to incorportate these factors in project-evaluation criteria.

Many states will need to alter current practice to reflect performance measurement area emphases, particularly for the *System Reliability* and *Congestion Mitigation* national goal areas. The vast majority of state transportation agencies are not measuring delay and throughput based on person-based metrics, as is codified in Federal Regulations part 490.507(a) of the Federal Register (*24*). Most states indicated that they do not have ways to evaluate all projects under the same congestion-based indicators, and instead rely on separate project-level analyses, which limit the ability to select projects that most cost-effectively meet the performance area specified.

Federal programs can be realigned to better support state's progress toward developing more strategic, multimodal, and performance and outcome-based approaches. Federal programs remain largely restrictive when it comes to project types enforced by formula-based requirements. States wishing to be more strategic have to work within the current constraints of federal funding programs. This can hamper their efforts toward creating a more multimodal and integrated transportation system. From the survey findings, more states would be more supportive of freight funding created under the FAST Act if there was not the restriction on non-highway projects comprising up to 10% of total NHFP and INFRA funds (*20*).

States also say that current funding is insufficient to meet their multimodal planning and assessment needs. Most indicated that support of the assessment and implementation of intermodal and multimodal projects, and routine non-NHS road projects lacks emphasis in federal funding. The only current multimodal and multi-state jurisdictional funding source at the federal level is the TIGER Discretionary Grants program (*26*). Formula funds comprise a far greater amount of the funding distributed to states, but lack the necessary flexibility. Several states apply for very limited TIGER funding to obtain funding flexibility that states cannot obtain through formula funds. Surface transportation block grant funds are arguably the most flexible formula funding source, but the survey indicates that outside of bicycle and pedestrian projects most of the funds serve traditional purposes, not the types of projects that are otherwise difficult to undertake.

Formula funds comprise most of the transportation funding that states receive from the federal government (*27*), but these funds offer less incentive for states to enhance performance, ensure infrastructure quality, and address safety goals than the use of competitive funds or formula funds that are tied to states making substantial progress or estimating future performance through evidence-based, long-term strategic investment decisions.

One significant predicament in relying so heavily on formula funds is that states may tend toward using traditional processes at a time when performance-based processes are the ones being sought at the federal level. Formula funds are only effective when projects comprise the highest and best use of those funds. When that is not the case, system performance can be less optimal than could otherwise be attained.

Conclusion

Our research findings indicate states continue to be in the process of adopting performance linkages to plans and programs. Within the context of national performance requirements, we recommend that states move toward emphasizing consistent use of evaluation criteria across capital programs in order to describe the degree to which targets are achieved through projects that go through the programming process. Project-evaluation methods based on multi-criteria decision analysis appear to be of growing interest to state DOTs that are adopting performance-based plans and programs. Multi-criteria methods can enable states to identify how specific projects help to achieve targets and align with transportation performance goals.

This research offers an overall assessment, meaning further and more detailed analysis is necessary to show how DOTs can move forward, given their diversity of approaches. However, from the overall review there are gaps that remain in adopting performance-based planning and programming. Multimodal focuses are even more lacking, due to multimodal planning deficiencies that are then exacerbated by modal funding barriers in project evaluation and selection. Modal biases in funding and planning can hamper performance progress and achievement. Given states continue to lack evaluation tools for mode-neutral project assessment, future research should examine the range of methods for calculating person-based and ton-based movement in project evaluation, and to advance their use for informing project selection. Another significant finding is the growth in interest in accessibility metrics. Further tools and methods can be deployed in project-evaluation processes to account for increases in access to vital destinations from individual capital expansion projects. As states advance performance management practices, project-evaluation methods that better align with national performance goal areas merit greater attention.

Author Contributions

The authors confirm contribution to the paper as follows: study conception and design: CR, ES; data collection: ES; analysis and interpretation of results: ES, CR; draft manuscript preparation: ES, CR. All authors reviewed the results and approved the final version of the manuscript.

References

1. Pei, Y. L., J. M. Fischer, and A. A. Amekudzi. *Performance Measurement in State Departments of Transportation: A Literature Review and Survey of Current Practice*. Georgia Institute of Technology, 2010. http://www.irg.ce.gatech.edu/sites/default/files/files/pei_fischer_ amekudzi.pdf.

2. Government Accountability Office. *Statewide Transportation Planning: Opportunities Exist to Transition to Performance-Based Planning and Federal Oversight*. Publication GAO-11-17. U.S. Government Accountability Office, 2010.
3. Gunasekera, K., and I. Hirschman. *Cross Mode Project Prioritization*. Publication NCHRP Project 08-36, Task 112. AASHTO, 2014.
4. Sinha, K. C., and S. Labi. *Transportation Decision Making: Principles of Project Evaluation and Programming*. John Wiley & Sons, Inc., 2007.
5. Maggiore, M., and K. M. Ford. *Guide to Cross-Asset Resource Allocation and the Impact on Transportation System Performance*. Publication. NCHRP Report 591. CH2M Hill and High Street Consulting Group, 2015.
6. Grant, M., J. D'Ignazio, A. Bond, and A. McKeeman. *Performance-based Planning and Programming Guidebook*. Publication. FHWA-HEP-13-041. U.S. DOT FHWA, 2013.
7. Federal Highway Administration. *Statewide & Nonmetropolitan Transportation Planning Final Rule: External FHWA/FTA Webinar for Stakeholders*. USDOT, June 14, 2016b. https://www.fhwa.dot.gov/tpm/rule/160614presentation.pdf.
8. Federal Highway Administration. *Fixing America's Surface Transportation Act or "FAST Act" A Summary of Highway Provisions*. U.S. Department of Transportation Federal Highway Administration, 2016a.
9. Congress.Gov. *H.R.22 Fixing America's Surface Transportation Act (FAST Act)*. Public Law No: 114-94, 2015. https://www.congress.gov/114/bills/hr22/BILLS-114hr22enr.pdf.
10. Government Publishing Office. *Notice of Funding Opportunity for the Department of Transportation's Nationally Significant Freight and Highway Projects (INFRA Grants) for Fiscal Years 2017 and 2018*. Publication. USGAO Federal Register Vol. 82, No. 127, July 5, 2017.
11. Fischer, J. M. *Transportation Performance Management for Livability and Social Sustainability: Developing and Applying a Conceptual Framework*. PhD dissertation. Georgia Institute of Technology, Atlanta, 2014.
12. Brown, I., and R. Ginsburg. *Research Report: Proposal for a Program and Project Prioritization Framework*. Publication, Urban Transportation Center, 2016.
13. Meyer, M. D., and E. Miller. *Transportation Planning: A Decision Oriented Approach*, 2014.
14. Cambridge Systematics & HDR, Inc. *NCHRP Report 591: Factors that Support the Planning-programming Linkage*. Publication. Transportation Research Board of the National Academies, Washington, D.C., 2007.
15. Sonnenberg, A. H., F. Southworth, M. D. Meyer, and C. Comer. Statewide Multimodal Planning: Current Practice at State Departments of Transportation. *Transportation Research Record: Journal of the Transportation Research Board*, 2013. 2397: 1–10.
16. Anderson, G., R. Bellis, A. Dodds, M. Ebeling, A. Gad, A. Grodnik, B. Holloway, L. M. MacKinnon, T. Madrecki, C. McCahill, R. Millar, D. Moser, S. Munger, C. Spahr, E. Sundquist, and R. Webber. *The Innovative DOT: A Handbook of Policy and Practice*. Smart Growth America and State Smart Transportation Initiative, 2014.
17. *Chicago Metropolitan Agency for Planning Peer Exchange*. FHWA & FTA, U.S. Department of Transportation, 2012.
18. Barfod, M. B., and S. Leleur. *Multi-criteria Decision Analysis for Use in Transport Decision Making*. Technical University of Denmark, Transport, DTU Lyngby, 2014.
19. Parnell, G. S., P. J. Driscoll, and D. L. Henderson. *Decision Making in Systems Engineering and Management*. Wiley Series in Systems Engineering and Management, 2008.
20. Sperling, E. A. Advancing New Strategic Focuses Through Performance-based Evaluation – The Growth of State DOT Approaches. thesis. Georgia Institute of Technology, Atlanta, 2017.
21. ODOT. *Enhance Proposal Review Process: An Overview*. Oregon Department of Transportation, 2015.
22. UDOT. *Capacity Projects – Prioritization*. Utah Department of Transportation, July 18, 2016. https://www.udot.utah.gov/main/f?p=100%3Apg%3A0%3A%3A%3A1%3AT%2CV%3A2275%2C. Accessed June 10, 2017.
23. VDOT. *SMART SCALE Technical Guide: Prepared for Commonwealth Transportation Board*. Virginia Department of Transportation, 2016b.
24. GPO. *Federal Register: Rules and Regulations*. Vol. 82, No. 11, January 18, 2017. https://www.gpo.gov/fdsys/pkg/FR-2017-01-18/pdf/2017-00681.pdf
25. MaineDOT. Industrial Rail Access Program (IRAP) Application for Assistance. *Maine Department of Transportation*, 2017b.
26. USDOT. *TIGER Discretionary Grants*. U.S. Department of Transportation, 2017. https://www.transportation.gov/tiger.
27. FHWA. *Table 1, FHWA Notice Apportionment of Federal-Aid Highway Program Funds for Fiscal Year (FY) 2016*. Federal Highway Administration, 2016c.

The Standing Committee on Transportation Programming and Investment Decision-Making (ADA50) peer-reviewed this paper (18-03600).

Article

TRR
JOURNAL OF THE TRANSPORTATION RESEARCH BOARD

Transportation Research Record
2018, Vol. 2672(51) 79–87

DOI: 10.1177/0361198118777619
journals.sagepub.com/home/trr
SAGE

Profile Monitoring for Examining Impact of Hurricane Katrina and Rita on Highway Construction Cost

Minsoo Baek[1] and Baabak Ashuri[1,2]

Abstract

Extensive destruction and damage to communities in the aftermath of large-scale disasters such as hurricanes, floods, and earthquakes can cause substantial disorder in social and economic activity. The ensuing repercussions to social and economic systems lead to considerable uncertainty in continuing existing construction projects and implementing new reconstruction is problematic for accurately determining construction costs and making appropriate investment decisions. A large-scale disaster magnifies the importance of empirically examining variations in construction costs. Therefore, the overarching objective of this research is to analyze submitted unit price bids for superpave asphaltic concrete line items for highway projects let in the state of Louisiana after hurricanes Katrina and Rita. We employ two-way analysis of variance and a cumulative sum (CUSUM) control chart to monitor and detect changes in submitted unit price bids for hurricane- and non-hurricane-impacted areas. The results of the CUSUM control chart indicate that unit price bids for asphalt line items in the impacted areas significantly increased immediately after Hurricane Katrina and Rita hit the Gulf of Mexico, and those in the non-hurricane-impacted areas significantly increased six months after the hurricane. The findings of this research enable transportation agencies to more closely estimate the cost of rebuilding highway projects and make more-informed investment decisions for highway recovery programs in the aftermath of large-scale disasters.

Hurricanes Katrina and Rita hit the Gulf of Mexico coast in August and September 2005, respectively, and devastated a great deal of property in Louisiana, Mississippi, and Alabama. According to the Department of Homeland Security (DHS), the storm surge and flooding caused by Hurricane Katrina caused 1,326 deaths, displaced more than 700,000 people from the Gulf Coast regions, 273,000 of whom were evacuated to shelters, and destroyed approximately 300,000 homes (*1*). Such large-scale disasters cause extensive destruction and damage to the infrastructure and buildings (*2*) and disorder in social and economic activity, both of which impose a significant burden on entities responsible for implementing construction and reconstruction projects. In the aftermath of a large-scale disaster, an unprecedented increase in the demand for construction and the disruption of transportation and resource supply systems generate considerable uncertainty that leads to inaccurate cost estimates for construction projects and inappropriate investment decisions (*3*, *4*). In addition, construction projects under such uncertainty often experience cost overruns, financial problems, and project delays or cancellations (*5*). These problems call for research that improves our understanding of changes in construction costs after a natural disaster.

Several studies have investigated the impact of natural disasters on the construction industry. For example, a study by Guimaraes et al. examined changes in economic activity in the aftermath of Hurricane Hugo, which struck the U.S coast in South Carolina in 1989 (*6*). The authors found that construction income significantly increased after the hurricane and suggested that the main causes of the increase were greater regional activity in construction sectors (e.g., residents and businesses) and greater financial flow in the form of disaster relief and insurance claims to communities. In another study, Chang et al. found that construction agencies often face difficulty procuring skilled construction labor and

[1]School of Building Construction, Georgia Institute of Technology, Atlanta, GA
[2]School of Civil & Environmental Engineering, Georgia Institute of Technology, Atlanta, GA

Corresponding Author:
Address correspondence to Minsoo Baek: minsoo.baek@gatech.edu

materials for implementing reconstruction projects after large-scale disasters (7). The resource availability for post-disaster reconstruction can cause significant cost changes for reconstruction projects. From a survey of building contractors, the authors identified influential factors (e.g., quantity of resources required, competency of resourcing manager and qualification of contractors, general economic environment, and resource transportation cost) that affected resource availability for housing recovery construction after the 2008 Wenchuan earthquake in China.

To explain construction cost changes for residential and commercial properties after the 2002 to 2010 hurricane seasons in Florida and the other states on the Atlantic and Gulf of Mexico coasts, Olsen and Porter developed an exploratory model that found that these regions underwent significant cost changes after large storms (8). The authors concluded that labor costs were a major driver of construction cost changes for repairing residential and commercial properties. Cheng and Wilmot studied the impact of Hurricanes Katrina and Rita on the Louisiana highway construction cost index, consisting of six major bid items (9). They also concluded that the index in the hurricane-impacted area significantly increased after the hurricanes and declined during the subsequent two quarters while the index in the non-hurricane-impacted area declined after the hurricane. The studies mentioned above have not examined the effects of natural disasters on the costs of highway construction, which suggests the need for an empirical analysis of such costs.

Research Objective

The overarching objective of this research is to analyze the impact of Hurricane Katrina and Rita on submitted unit price bids for superpave asphaltic concrete line items. To do so, this paper has the following sub-objectives:

1. To monitor the changes in submitted unit price bids for hurricane-impacted and non-hurricane-impacted areas.
2. To detect process shifts in submitted unit prices after Hurricanes Katrina and Rita.
3. To investigate the long-term impact of Hurricanes Katrina and Rita on variations among submitted unit prices.

Hurricane-Impacted Areas

Figure 1 shows the 64 parishes in Louisiana, highlighted in gray, devastated by Hurricanes Katrina and Rita. The federal government declared these parishes, listed on the Federal Emergency Management Agency (FEMA) official website to be disaster areas (*10*, *11*). Hurricane Katrina impacted nine parishes, including Jefferson, Orleans, Plaquemines, St. Bernard, St. Charles, St. John the Baptist, St. Tammany, and Tangipahoa in the southeastern region in the state of Louisiana, and Hurricane Rita also impacted nine parishes, including Calcasieu, Cameron, Iberia, Lafourche, St. Charles, St. John the Baptist, St. Mary, Terrebonne, and Vermilion in the southern regions in the state. Thus, focusing on these parishes, this paper monitors changes in submitted unit price bids for highway construction projects after Hurricane Katrina and Rita.

Research Methodology

This study employs two-way analysis of variance (ANOVA) and Tukey–Kramer tests to compare the mean differences between hurricane- and non-hurricane-

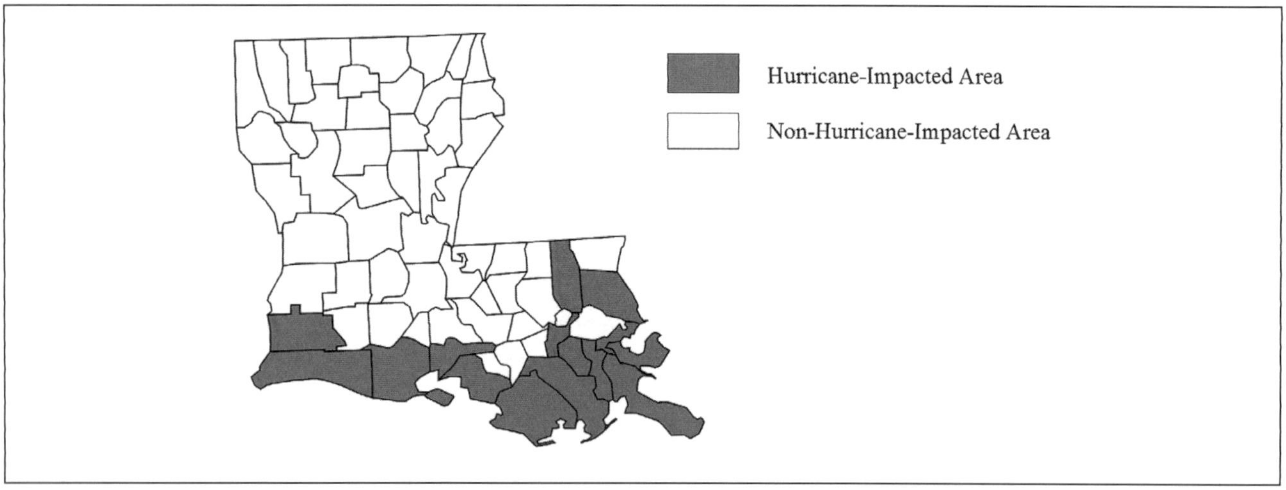

Figure 1. Parishes impacted by Hurricanes Katrina and Rita.

impacted areas before and after Hurricanes Katrina and Rita. For the comparison, this paper divides the dataset into five periods: the period before the hurricanes, the first quarter, the second quarter, the third quarter, and the fourth quarter after the hurricanes, and the period after the fourth quarter. The null hypothesis of the two-way ANOVA indicates that the means of all periods are equal (H_0:$\mu_a = \mu_i, i = b\ to\ e$). As each period has an unequal sample size, the Tukey–Kramer method is an effective tool for conducting multiple comparisons of means. Applying this test, we conduct pairwise comparisons between the period during which Hurricanes Karina and Rita occurred and subsequent periods.

We adopt statistical process control (SPC) for monitoring and detecting changes in submitted unit price bids after Hurricanes Katrina and Rita in the state of Louisiana. SPC is a statistical method for the quality control of a process. For this study, we selected a cumulative sum (CUSUM) control chart, one of the tools of SPC, to monitor and detect changes in submitted unit price bids. CUSUM is a powerful tool for detecting a small shift in a process (*12*).

SPC techniques have been developed in practical situations in which a process/product is functionally dependent on one or more explanatory variables. In such cases, a profile monitoring technique that takes into account functional relationships between the process and explanatory variables is applicable to the monitoring process (*12*). A profile-monitoring technique uses residuals from a developed regression model to monitor the process. In addition, monitoring regression residuals with the CUSUM control chart instead of using original data values enhances sensitivity to small shifts (*12*). Thus, this paper applies CUSUM control charts to regression residuals for hurricane-impacted and non-hurricane-impacted areas and investigates the impact of Hurricanes Katrina and Rita on variations among submitted unit price bids for superpave asphaltic concrete line items. Lastly, to monitor the long-term impact of Hurricanes Katrina and Rita, we used a dataset between 2004 and 2015 in our CUSUM control chart analysis, following the steps of the research methodology:

1. Comparing the mean differences in submitted unit price bids between the period before and after Hurricanes Katrina and Rita.
2. Developing regression models for the hurricane- and non-hurricane-impacted areas before the hurricanes hit the Gulf Coast.
3. Examining the relationships between submitted unit price bids and explanatory variables.
4. Computing residuals by subtracting the submitted unit price bids from the predicted values for the entire dataset (i.e., the dataset before and after Hurricanes Katrina and Rita).
5. Applying CUSUM control charts to the residuals for both areas.
6. Diagnosing CUSUM control charts to detect any shifts in the regression residuals and out-of-control signals after Hurricanes Katrina and Rita for the hurricane- and non-hurricane-impacted areas.
7. Investigating the long-term impact of Hurricanes Katrina and Rita on the variability of the submitted unit price bids.

Dataset

As superpave asphaltic concrete is one of most common asphalt line items for pavement projects in the State of Louisiana, the dataset consists of submitted unit price bids for this line item retrieved from the BidTabs database of Oman Systems. This study monitors unit price bids submitted between 2004 and 2015 for highway pavement projects let in the state and refers to submitted unit price bids by construction contractors between 2004 and 2008 to analyze the short-term impact of Hurricanes Katrina and Rita. To investigate the long-term impact of the hurricanes, this study analyzes the submitted unit price bids for superpave asphaltic concrete between 2004 and 2015. In addition, it defines the period between August 29, 2005, and September 18, 2005 as the time during which Hurricanes Katrina and Rita hit the Gulf Coast of the United States.

To develop regression models for both hurricane- and non-hurricane-impacted areas, we selected the explanatory variables that most strongly reflect project characteristics and construction market conditions:

- Quantity of the bid item: the volume of superpave asphaltic concrete line items in the submitted unit price bid.
- Total bid price: the lowest total bid price submitted by highway contractors who bid on a project.
- Unemployment: the number of laborers who are eligible to work but unable to find a job, measured at the county level in the state of Louisiana.
- Building cost index: the composite index that measures the overall performance of the building construction industry over time by tracking local prices of skilled labor and materials collected in the city of New Orleans, Louisiana.
- Building permits for new residential construction: the number of new housing units in the southern region authorized by building permits for privately-owned residential construction.

Table 1. Results of the Two-way ANOVA Analysis for the Hurricane-Impacted and Non-Hurricane-Impacted Areas

	Hurricane-impacted area					Non-hurricane-impacted area				
Source	Sum of squares	df	Mean square	F	Sig.	Sum of squares	df	Mean square	F	Sig.
Model	53676.31	5	10735.26	5.63	0.000	149615.01	5	29923.01	5.63	0.000
Error	331827.51	174	1907.06			1350851.38	466	2898.82		
Corrected total	385503.82	179				1500466.39	471			

Note: Sig. = significance level (*P*-value); df = degree of freedom; F = F statistic.

Table 2. Results of Tukey-Kramer Multiple Comparisons for the Hurricane-Impacted and Non-Hurricane-Impacted Areas

	Hurricane-impacted area				Non-hurricane-impacted area			
	$\lvert\mu_{BF}-\mu_i\rvert$	Sig.	95% confidence limits		$\lvert\mu_{BF}-\mu_i\rvert$	Sig.	95% confidence limits	
BF and Q1	36.708	0.711	−37.427	110.842	8.262	0.980	−24.753	41.277
BF and Q2	13.595	0.881	−22.081	49.270	13.842	0.670	−12.722	40.407
BF and Q3	43.842	0.009	7.124	80.559	54.424	0.000	26.460	82.390
BF and Q4	30.755	0.099	−3.135	64.645	40.152	0.011	5.910	74.395
BF and AF	35.071	0.000	12.935	57.206	31.637	0.000	15.044	48.231

Note: Sig. = significance level (*P*-value); BF = the period before Hurricanes Katrina and Rita, Q1 = 1st quarter, between September and November 2005; Q2 = 2nd quarter, between December 2005 and February, 2006; Q3 = 3rd quarter, between March and May 2006; Q4 = 4th quarter, between June and August 2006; AF = the period between September 2006 and December 2007; and $\lvert\mu_{BF}-\mu_i\rvert$= the difference between the means (*I* = Q1, Q2, Q3, Q4, and AF).

Results

Short-Term Impact of Hurricanes Katrina and Rita

Table 1 lists the results of the two-way ANOVA analysis for the hurricane- and non-hurricane-impacted areas. The results indicate that the means of the five periods before and after the hurricanes statistically differed by rejecting the null hypothesis at a significance level (*P*-value) of 0.05. From these results, we conclude that the submitted unit price bids for the hurricane- and non-hurricane-impacted areas significantly differed in both areas after the hurricanes. To compare the means of multiple periods, we applied the Tukey–Kramer test. The results of the test for both the hurricane- and non-hurricane-impacted areas are listed in Table 2, which shows no statistical significant difference between a period before Hurricane Katrina and the first and second quarters after the hurricanes (P-value > 0.05). In the hurricane-impacted area, the Tukey–Kramer multiple comparisons revealed a significant mean difference between the period before the hurricanes and the third quarter and the last period between September 2006 and December 2007 after the hurricanes (P-value $= 0.009$ and 0.001, respectively). In the non-hurricane-impacted area, however, the submitted unit price bids in the third and fourth quarters and the period between September 2006 and December 2007 exhibited significant mean differences.

From these results, we conclude that a significant cost change for both the hurricane- and non-hurricane-impacted areas occurred in the third quarter between March and May 2006. Results of the two-way ANOVA and Tukey–Kramer tests showed significant cost changes during a period before and after Hurricanes Katrina and Rita.

We applied ordinary least squares regression modeling to develop regression models with the submitted bid prices and explanatory variables. To determine the linear relationships between the submitted unit price bids and explanatory variables, we conducted a scatter plot assessment and Pearson correlation tests. We used natural logarithms to transform the quantity of bid items among the explanatory variables. Tables 3 and 4 list the results of the regression analysis using the submitted unit price bids before Hurricanes Katrina and Rita and project-related and construction market variables for the hurricane- and non-hurricane-impacted areas. From the results of the regression model of the hurricane-impacted area, the quantity of bid items and level of unemployment exhibit a strong relationship with the submitted unit price bids at a significance level (*P*-value) of 0.05 while holding all other variables constant. From the results of the regression model of the non-hurricane-impacted area, the quantity of bid items and the total bid price significantly impact the submitted unit price bids at a significance level (*P*-value) of 0.05 while holding all other variables constant. The quantity of bid items from both regression models exhibits a negative relationship with submitted unit price bids. Variance inflation

Table 3. Results of the Regression Analysis for the Hurricane- and Non-Hurricane-Impacted Areas

	Hurricane-impacted area			Non-hurricane-impacted area		
Model	Coefficients	t (Sig.)	VIF	Coefficients	t (Sig.)	VIF
Constant	216.259	2.904 (0.005)		237.336	3.358 (0.001)	
Quantity of bid item (transformed in natural logarithm form)	−8.621	−6.523 (0.000)	1.209	−18.763	−20.165 (0.000)	1.148
Total bid price	2.284×10^{-7}	1.118 (0.267)	1.280	1.486×10^{-6}	4.526 (0.000)	1.120
Unemployment	0.002	2.888 (0.005)	1.084	−0.001	−1.791 (0.075)	1.024
Crude oil price	0.491	0.878 (0.383)	4.029	0.384	0.814 (0.417)	4.300
Building cost index, New Orleans	−0.024	−0.964 (0.338)	2.865	−0.015	−0.659 (0.511)	2.891
Building permits for new residential construction, south	−0.040	−0.641 (0.524)	2.276	0.015	0.287 (0.774)	2.241
	Adjusted R squared = 0.415			Adjusted R squared = 0.711		

Note: VIF = variance inflation factors; Sig. = significance level (*P*-value).

Table 4. ANOVA Test for the Regression Models for the Hurricane- and Non-Hurricane-Impacted Areas

	Hurricane-impacted area					Non-hurricane-impacted area				
Model	Sum of squares	df	Mean square	F	Sig.	Sum of squares	df	Mean square	F	Sig.
Regression	16042.572	6	2673.762	9.509	0.000	179172.326	6	29862.054	70.024	0.000
Residual	18558.361	66	281.187			69085.521	162	426.454		
Total	34600.933	72				248257.848	168			

Note: Sig. = significance level (*P*-value); df = degree of freedom; F = F statistic.

factors (VIFs) are less than 10. Therefore, the model has no multicollinearity issue. The results of the ANOVA for the regression models, listed in Table 4, indicate that the regression models for both the hurricane- and non-hurricane-impacted areas statistically significant.

We calculated regression residuals for the entire dataset before and after Hurricane Katrina and Rita by subtracting the submitted unit price bids from the predicted values, computed by the estimated regression equations. Then, to monitor the process of the submitted unit price bids, we applied the CUSUM control chart to the regression residuals. CUSUM control charts can be created by plotting the cumulative sums of the deviations of the residuals from the target value (μ_0). Creating a CUSUM control chart requires the construction of a tabular CUSUM, which entails the use of one-sided upper and lower CUSUMs, denoted by C^+ and C^-, which are computed by the following equations (*12*):

$$C_i^+ = \max\left[0, x_i - (\mu_0 + K) + C_{i-1}^+\right] \quad (1)$$

$$C_i^- = \max\left[0, (\mu_0 - K) - x_i + C_{i-1}^-\right] \quad (2)$$

In which the starting values are $C_0^+ = C_0^- = 0$, μ_0 (target value) is the in-control values/residuals of the mean before Hurricanes Katrina and Rita, and K is the reference value, calculated at $K = |\mu_1 - \mu_0|/2$, where μ_1 is the out-of-control values/residuals of the mean after Hurricanes Katrina and Rita. While the upper CUSUM, C_i^+, detects an increase in the submitted unit price bids for asphalt line items, the lower CUSUM, C_i^-, detects a decrease in the submitted unit price bids. In addition, the CUSUM control chart consists of upper and lower control limits, which are estimated by decision interval H. The upper and lower control limits are defined as follows (*12*):

$$Upper\ Control\ Limit\ (UCL) = H\sigma \quad (3)$$

$$Lower\ Control\ Limit(LCL) = -H\sigma \quad (4)$$

where a reasonable value for H is 5 and σ is the standard deviation of the in-control values. Thus, the upper and lower control limits for the hurricane- and non-hurricane-impacted areas were ± 80.274 and ± 101.393, respectively. If C_i^+ and C_i^- exceed both upper and lower control limits, the process is defined as an out of control state, indicating that a significant increase or decrease in the submitted unit price bids. Determining the recovery period requires the number of consecutive periods in which CUSUM C_i^+ or C_i^- have been nonzero prior to the first period that is out of state after Hurricanes Katrina and Rita (*12*).

Figures 2 and 3 plot the CUSUM from calculations in Equations (1), (2), (3), and (4), mentioned above.

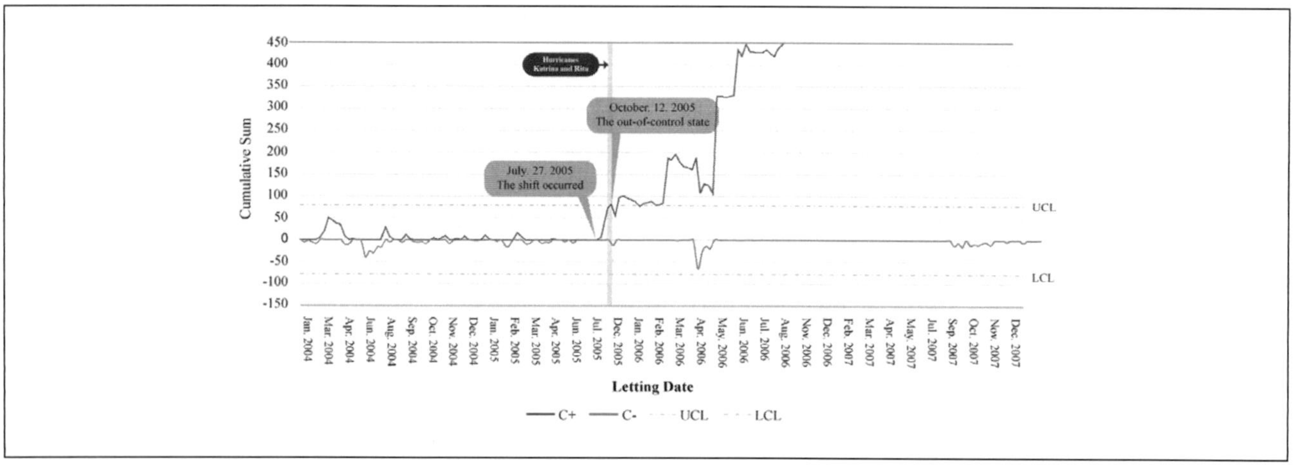

Figure 2. CUSUM control chart for the hurricane-impacted area.

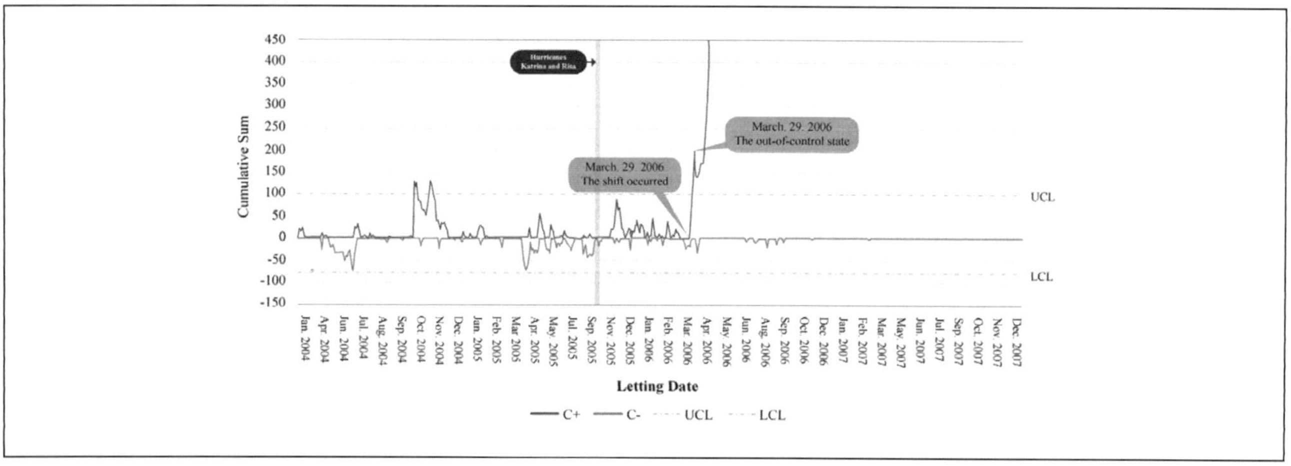

Figure 3. CUSUM control chart for the non-hurricane-impacted area.

Figure 2 presents the results of the CUSUM control chart for the hurricane-impacted area, which showed a significant increase in the submitted unit price bids for superpave asphaltic concrete line items detected immediately after Hurricanes Katrina and Rita. A submitted unit price bid for a project let in October 2005 was the first observation in the out-of-control state after Hurricanes Katrina and Rita. In addition, the submitted unit price bids in the hurricane-impacted area started to shift in July 2005. According to the third quarterly cost report of *Engineering News Report* (ENR), a boom in cleanup and reconstruction activity in the hurricane-impacted area immediately after Hurricanes Katrina and Rita caused a shortage of materials and skilled labor that led to price increases in construction resources (*13*). In addition, the severe supply disruptions caused by the shutdown of the energy infrastructure (e.g., oil refineries), cement plants, and transport systems such as rails, roads, and bridges for construction resources, along with the damage they sustained in the aftermath of Hurricanes Katrina and Rita resulted in significant cost increases in construction materials, which led to higher bids for construction projects (*14*). The recovery period was not identifiable in the analysis of the short-term impact of Hurricanes Katrina and Rita.

Figure 3 shows the CUSUM control chart for the non-hurricane-impacted area. Although significant variability in the submitted unit price bids occurred in the aftermath of Hurricanes Katrina and Rita, none of the CUSUM C_i^+ or C_i^- were out of control until March 2006. The non-hurricane-impacted area experienced a shift in the submitted unit price bid in March 2006, when the first out-of-control state in the non-hurricane-impacted area also occurred. Thus, the

Table 5. Results of the Regression Analysis for Submitted Unit Price Bids between 2004 and 2015

Model	Coefficients	t (Sig.)	VIF
Constant	195.796	14.437 (0.000)	
Quantity of bid item (transformed in natural logarithm form)	−19.474	−52.501 (0.000)	1.022
Total bid price	3.790×10^{-7}	7.193 (0.000)	1.028
Unemployment	3.951×10^{-4}	2.029 (0.042)	1.024
Crude oil price	0.109	2.751 (0.006)	1.306
Building cost index, New Orleans	0.019	5.443 (0.000)	1.680
Building permits for new residential construction, south	−0.330	−7.344 (0.000)	1.683
	Adjusted R squared= 0.456		

Note: VIF = variance inflation factors; Sig. = significance level (*P*-value).

Table 6. ANOVA Test for the Regression Models for the Submitted Unit Price Bids between 2004 and 2015

Model	Sum of squares	df	Mean square	F	Sig.
Regression	6773715.917	6	1128952.653	534.342	0.000
Residual	8045509.112	3808	2112.791		
Total	14819225.03	3814			

Note: Sig. = significance level (*P*-value); df = degree of freedom; F = F statistic.

non-hurricane-impacted area experienced a significant increase in submitted unit price bids during the six months following Hurricanes Katrina and Rita.

Figures 2 and 3 illustrate a major shift between March and July 2006 that occurs in both the hurricane- and non-hurricane-impacted areas. This finding is consistent with the results of the Tukey–Kramer test above. The major cause of this shift is a boom in reconstruction activity. A year after Hurricanes Katrina and Rita hit the Gulf of Mexico, more than 38,000 building permits were issued for rebuilding residential property (*15*). Moreover, approximately \$5 billion in federal grants went to the state of Louisiana for aid in rebuilding damaged housing and other infrastructures in May 2006 (*16*). The federal support and the massive building permits that prompted a boom in reconstruction activities led to increased demand for construction services, which in turn, caused higher bids for higher projects let in the entire state of Louisiana. The CUSUM control chart for non-hurricane-impacted areas was not able to detect the recovery period in the short-term analysis of Hurricanes Katrina and Rita.

Long-Term Impact of Hurricane Katrina and Rita

We use the CUSUM control chart to diagnose the long-term impact of Hurricanes Katrina and Rita. We begin by developing a regression model using submitted unit price bids for highway projects let throughout the state of Louisiana between 2004 and 2015 with the key explanatory variables. The results of the regression model appear in Tables 5 and 6. The identified variables are able to explain variations among the submitted unit price bids for superpave asphaltic concrete with a 95% significance level (P-value < 0.05). In addition, the ANOVA test concluded that the developed model was statistically significant at a 95% confidence (P-value < 0.05).

After accounting for the variability explained by the identified variables in the regression model, we use the residuals from the developed regression model to develop a CUSUM control chart for measuring the variability among the submitted unit price bids, shown in Figure 4. Five standard deviations (5 sigma) are depicted in Figure 4 as the upper- and lower-control limits (i.e., 229.64 and −229.64, respectively) to identify the significance level of the variability. The first out-of-control signal on the CUSUM control chart is detected on April 26, 2006, and then a shift likely occurs after Hurricane Katrina and Rita between February 22, 2006 and March 29, 2006. In addition, significant variability in the submitted unit price between March 2006 and December 2006 is noticeable. The U.S. Department of Housing and Urban Development (HUD) approved \$368.4 million to fund rebuilding plans that would meet the infrastructure needs of the state and \$10.8 billion to fund Louisiana's Road Home Program (*16*). The rebuilding plans led to a significant increase in construction demand and a shortage of construction resources (e.g., materials, workers, equipment) in the state, which may have resulted in the significant variability among submitted unit price bids. Furthermore, according to a study conducted by Kates et al., the period of the restoration process for electricity,

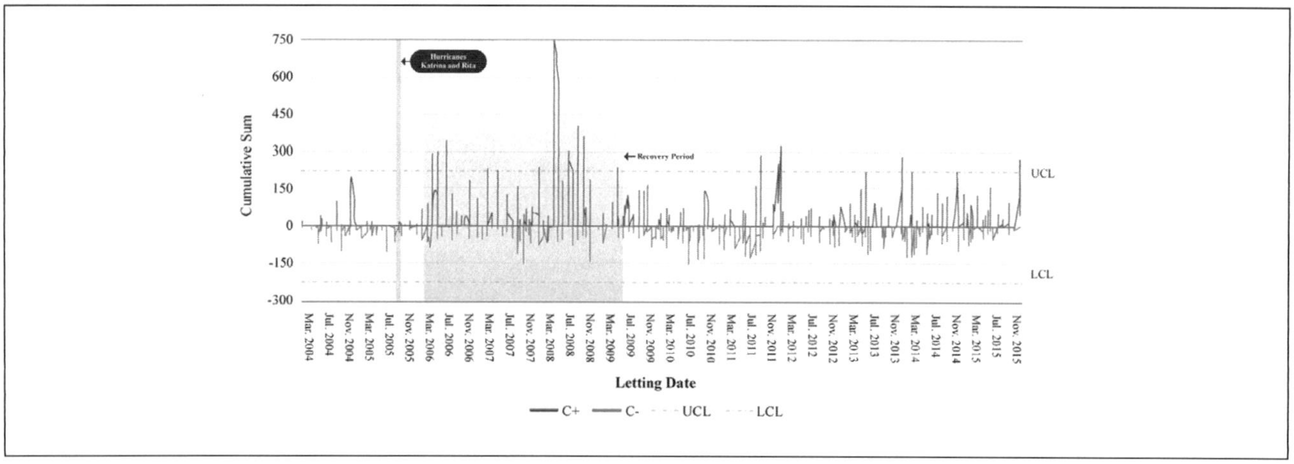

Figure 4. CUSUM control chart for submitted unit price bids in the state of Louisiana between 2004 and 2015.

gas, public transportation, schools, hospitals, and food stores was 40 weeks (until June of 2006) (*17*). After the completion of the restoration process, the reconstruction process gained momentum in the reestablishment of the infrastructure, housing, and jobs for the destroyed areas (*17*). During the Atlantic hurricane season in 2008, Louisiana experienced two devastating hurricanes, Hurricanes Gustave (Category 4) and Ike (Category 4), which may have caused additional variability in the submitted unit price bids. After the great recession of 2008 and 2009, the variability subsides with unit price bids returning to their normal/pre-hurricane levels between June 2009 and August 2011. Therefore, the recovery period for the submitted unit price bids after Hurricanes Katrina and Rita occurred between February 2006 and June 2009, approximately 170 weeks.

Conclusion

Hurricanes Katrina and Rita significantly impacted submitted unit price bids for superpave asphaltic concrete line items used on highway pavement projects. Two-way ANOVA and a Tukey–Kramer test showed statistically significant mean differences between the period before and after Hurricanes Katrina and Rita. In this study, we used a profile monitoring technique to investigate short- and long-term effects of these hurricanes on submitted unit price bids. In a short-term analysis, we detected significant shifts in the submitted unit price bids before and after the hurricanes for both hurricane- and non-hurricane-impacted areas. Results of a CUSUM control chart indicated that after Hurricanes Katrina and Rita hit the Gulf of Mexico, unit price bids for asphalt line items in the hurricane-impacted area significantly increased immediately after the hurricanes, and those in the non-hurricane-impacted area significantly increased six months after the hurricane. In addition, the results of both Tukey–Kramer multiple comparisons and CUSUM control charts indicated significant cost increases in submitted unit price bids in the third quarter, between March and May 2006 after Hurricanes Katrina and Rita. Furthermore, we investigated the long-term impact of these hurricanes on the submitted unit price bids between 2004 and 2015 and concluded that the variability among submitted unit price bids subsided; that is, bids returned to their normal levels in June 2009. We expect that these findings will help estimators and capital project planners set and adjust the costs of their highway construction projects and the timing of project letting while avoiding increases in construction demand and disruptions in transportation and resource supply systems during recovery and reconstruction in the aftermath of large-scale disasters.

Transportation agencies should ensure that post-disaster reconstruction, as well as new construction, is placed on safe ground and built on time and within budget. The results of this study will enable both construction agencies to improve their understanding of variations in construction costs after large-scale disasters and transportation agencies to more accurately estimate the cost of rebuilding highway projects. They should also help them make more-informed investment decisions for highway recovery programs in the aftermath of large-scale disasters. We expect that the proposed framework can be generalized to the monitoring of other line items after large-scale disasters.

Acknowledgments

The authors would like to thank John A. Oman, President of Oman Systems, Inc., for providing the data. The authors also extend their thanks and appreciation to the anonymous reviewers and the editor whose constructive suggestions greatly improved this work. This research is partially sponsored by the National Science Foundation (NSF) under grant number 1441208.

Author Contributions

The authors confirm contributions to the paper as follows: study conception and design: Minsoo Baek and Dr. Baabak Ashuri; data collection: Minsoo Baek; analysis and interpretation of results: Minsoo Baek and Dr. Baabak Ashuri; draft manuscript preparation: Minsoo Baek and Dr. Baabak Ashuri. All authors reviewed the results and approved the final version of the manuscript.

References

1. Department of Homeland Security. Office of Inspections and Special Reviews. *Performance Review of FEMA's Disaster Management Activities in Response to Hurricane Katrina*. Department of Homeland Security, Office of Inspector General, Office of Inspections and Special Reviews, Washington, D.C., 2006.
2. Hallegatte, S., and V. Przyluski. *The Economics of Natural Hazards: Concepts and Methods*. Policy Research Working Paper 5507. The World Bank Sustainable Development Network, Washington D.C., 2010.
3. Mendell, B. C. Benchmarking Construction Cost Changes in the U.S. South. *American Reconstruction*, 2006, pp. 1–6. http://forisk.com/wordpress//wp-content/assets/Construction-Index-final.pdf.
4. Senter, M. S. *Hurricane Katrina and Its Potential Impact on the Construction Industry*. Cooper and Scully, P.C., Dallas, Tex., 2006.
5. Peng, M. *Item-Level Quantity-based Preliminary Cost Estimating System for Highway Structures and Miscellaneous Construction*. The University of Texas at Austin, 2006.
6. Guimaraes, P., F. L. Hefner, and D. P. Woodward. Wealth and Income Effects of Natural Disasters: An Econometric Analysis of Hurricane Hugo. *The Review of Regional Studies*, Vol. 23, No. 2, 1993, pp. 97–114.
7. Chang, Y., S. Wilkinson, R. Potangaroa, and E. Seville. Identifying Factors Affecting Resource Availability for Post-Disaster Reconstruction: A Case Study in China. *Construction Management and Economics*, Vol. 29, No. 1, 2011, pp. 37–48.
8. Olsen, A. H., and K. A. Porter. Storm Surge to Demand Surge: Exploratory Study of Hurricanes, Labor Wages, and Material Prices. *Natural Hazards Review*, Vol. 14, No. 4, 2013, 247–257.
9. Cheng, G., and C. G. Wilmot. Louisiana Highway Construction Cost Trend after Hurricanes Katrina and Rita. *Journal of Construction Engineering and Management*, Vol. 135, No. 7, 2009, pp. 594–600.
10. Federal Emergency Management Agency. *Hurricane Katrina Flood Recovery Maps*. Federal Emergency Management Agency, 2015. http://www.fema.gov/hazard/flood/recoverydata/katrina/katrina_ms_methods.shtm. Accessed July 28, 2017.
11. Federal Emergency Management Agency. *Hurricane Rita Flood Recovery Maps*. Federal Emergency Management Agency, 2015. https://www.fema.gov/hurricane-rita-flood-recovery-maps-0. Accessed July 28, 2017.
12. Montgomery, D. C. *Introduction to Statistical Quality Control*. John Wiley & Sons, New York, 2009.
13. Engineering News-Record. *Third Quarterly Cost Report*. Engineering News-Record, Q3, 1–3, 2005.
14. Kowal, J., A. Lombardozzi, W. Snyders, and J. Weinhagen. *Price Highlights 2005: Higher Energy Prices Again Dominate Producer Prices*. U.S. Bureau of Labor Statistics, Washington, D.C., 2006.
15. Liu, A., M. Fellowes, and M. Mabanta. *Special Edition of the Katrina Index: A One-Year Review of Key Indicators of Recovery in Post-Storm New Orleans*. Brookings Institution, Washington, D.C., 2006.
16. U.S. Department of State. *Hurricane Katrina: What Government Is Doing?* 2006. http://www.state.gov/documents/organization/150082.pdf. Accessed July 28, 2017.
17. Kates, R. W., C. E. Colten, S. Laska, and S. P. Leatherman. Reconstruction of New Orleans after Hurricane Katrina: A Research Perspective. *Proceedings of the National Academy of Sciences*, Vol. 103, No. 40, 2006, pp. 14653–14660.

The Standing Committee on Transportation Programming and Investment Decision-Making (ADA50) peer-reviewed this paper (18-04568).

Article

TRR
JOURNAL OF THE TRANSPORTATION RESEARCH BOARD

Transportation Research Record
2018, Vol. 2672(51) 88–98

DOI: 10.1177/0361198118790378
journals.sagepub.com/home/trr
SAGE

Illustrating the Monetary Impact of Transportation Projects/Alternatives using the Values of Travel Time and Travel Time Reliability

Venkata R. Duddu[1], Srinivas S. Pulugurtha[1], and Praveena Penmetsa[2]

Abstract

State agencies, regional agencies, cities, towns, and local municipalities design and maintain transportation systems for the benefit of users by improving mobility, reducing travel time, and enhancing safety. Cost–benefit analysis based on travel time savings and the value of reliability helps these agencies in prioritizing transportation projects or when evaluating transportation alternatives. This paper illustrates the use of monetary values of travel time savings and travel time reliability, computed for the state of North Carolina, to help assess the impact of transportation projects or alternatives. The results obtained indicate that, based on the illustration of the effect and impact of various transportation projects or alternatives, both improved travel time and reliability on roads yield significant monetary benefits. However, from cost–benefit analysis, it is observed that greater benefits can be achieved through improved reliability compared with benefits from a decrease in travel time for a given section of road.

Travel demand has progressively increased with the development of modern civilization and the need for more travel over the past few decades. The subsequent effect of this increasing travel demand is recurring congestion on a limited road network, a growing air quality problem, and the lack of safe and reliable transportation systems. New roads are being constructed, existing roads are being widened, or alternate modes of transportation are being explored to reduce congestion and handle the growth. However, resource constraints (funding limitations) fueled by effective utilization of resources compel practitioners to assess the impact of transportation projects or alternatives before prioritization and implementation.

For years, it has been assumed that the fastest path from an origin to a destination is used by most motorists. These motorists usually plan for some expected delay caused by recurring congestion, which is common today in many urban areas. Further, motorists' approach toward trip planning seems to be changing because of fluctuations and uncertainty in traffic conditions. The unexpected or nonrecurring congestion caused by crashes, inclement weather, construction activity, and special events contributes most to these fluctuations and the uncertainty (*1–4*). Therefore, reliability (consistency or dependability in travel time) of a route is playing a more prominent role in departure and route choice decisions among various other travel time performance measures.

Reliability is defined as the probability that a component or system will perform a required function (without failure) for a given period of time, when used under stated operating conditions (*5*). For assessing travel time reliability or travel time variation, numerous indices were proposed by researchers in the past. The Federal Highway Administration recommends five standard measures of travel time reliability: 90th or 95th percentile travel time, travel time index (TTI), planning time index (PTI), buffer time index (BTI), and frequency that congestion exceeds some expected threshold (*6*). Lomax et al. grouped reliability related performance measures into three categories: statistical range, buffer time (BT) measures, and tardy trip indicators (*7*). The value of these performance measures and how they could be used

[1]Civil & Environmental Engineering Department/IDEAS Center, The University of North Carolina at Charlotte, Charlotte, NC
[2]Alabama Transportation Institute, The University of Alabama, Tuscaloosa, AL

Corresponding Author:
Address correspondence to Srinivas S. Pulugurtha: sspulugurtha@uncc.edu

to assess transportation projects or alternatives is vital as they are expected to be widely used in transportation planning, for project prioritization, and for allocation of resources.

The value of time refers to the monetary value travelers place on reducing their travel time. There exists a vast body of literature on the theoretical underpinning of the value of travel time (*8–11*), on travel time savings (*12*, *13*), and to a lesser extent on the value of reliability (*14–20*). Most of the past studies document limited and inconclusive research on the value of reliability, particularly based on motorists' perceptions.

Cost–benefit analysis helps state agencies, regional agencies, cities, towns, and local municipalities in prioritizing transportation projects/alternatives. It helps them efficiently and effectively use the limited available resources. In the cost–benefit analysis, the effect and impact of various transportation projects or alternatives are monetized to assess and make improved decisions. Currently, there exists limited documentation on the illustration and applicability of evaluating reliability, the value of travel time savings, and the value of reliability to assist with such decisions. Therefore, the objective of this paper is to illustrate the use and applicability of the value of travel time savings and the value of reliability to evaluate transportation projects or alternatives.

Monetary Values of Travel Time and Reliability

Pulugurtha et al. recently conducted a preliminary random general survey, followed by focus group survey and final random survey, to collect data and evaluate motorists' perceptions of reliability (*20*). The preliminary random general survey consisted of questions which were aimed at capturing stated preference choices to evaluate how the participants would react to hypothetical situations. Based on the results from preliminary random general survey of 417 participants, the average travel time for daily commute in the state of North Carolina is 23.47 min. The average additional trip time (planned maximum BT) is equal to 10.54 min. It does not seem to vary based on the average travel time. Some differences were observed in evaluation based on the income level of the participant. Low income group participants followed by high income group participants tend to plan for higher additional trip time than medium income participants. Males tend to choose an unreliable route more often than females. This indicates a difference in perception toward additional trip time by gender.

The generalized value of travel time for North Carolina is estimated to be $ 0.51 per minute. The generalized value of additional trip time is also $ 0.51, and participants are willing to pay, on average, $ 0.11 to reduce travel time by 1 min ($6.46 per hour). The value of additional trip time was observed to be different for peak and off-peak hours. It was also observed to vary by gender and income group of the participant. The computed value of additional trip time can be used as the maximum value of BT to evaluate the benefits of a proposed/designed transportation project/alternative.

Nine focus group meetings were carried out across North Carolina to collect data from 93 participants as a part of the research. The participants of the focus group survey were asked to choose between a reliable route and an unreliable route. The focus group survey aimed to observe participants' tradeoff between travel time savings and reliability. The results obtained from focus group survey indicate that higher percentages of participants are willing to opt for routes with a lower number of days of unreliability, as they will have some travel time savings. However, as the unreliability increases (number of days), participants start shifting from an unreliable route to a reliable route. The generalized value of BT for North Carolina based on focus group survey is estimated as $ 0.48 per minute. However, the focus group sample did not seem to be a good representation of the North Carolina population. Therefore, this was followed by the final random survey.

The final random survey comprised 357 participants who were asked to choose between two routes, a reliable route with longer travel time and an unreliable route with shorter travel time (as in the case of focus group survey). The maximum tolerance limit for unreliability is observed to be 4 days a month for a maximum BT of 20 min for a typical work day trip (~23 min).

Differences were observed when the findings from the final random survey were compared with the focus group survey. The generalized value of BT for North Carolina based on the final random survey is estimated as $ 0.45 per minute. This is based on the maximum BT that participants are willing to accept by selecting an unreliable route, hoping to save travel time for most of the trips, before choosing a reliable route. This value is less than the value of additional trip time from the preliminary random general survey and the value of BT from the focus group survey. It is recommended for use to evaluate the impact of transportation projects or alternatives, in addition to the generalized value of travel time. The monetary values of travel times and reliability estimated as a part of the research conducted for the North Carolina Department of Transportation (NCDOT) were used for illustration purposes in this paper.

Readers are encouraged to refer to Pulugurtha et al. for more details related to surveys conducted, sample, perceptions of participants, the value of reliability, and thresholds (*20*, *21*).

Value of Travel Time Savings: Illustration

The estimated generalized value of travel time savings for the state of North Carolina is \$6.46 per hour for every motorist. This value of travel time savings can be used to evaluate the benefits from a proposed/designed transportation project/alternative. First, the costs associated with the implementation of a transportation project/alternative are evaluated. Then, the benefits associated with the implementation of the transportation project/alternative (i.e., travel time savings in this case) are monetized. The costs and benefits are then compared to evaluate the breakeven period and overall returns. Along with the travel time savings, there could be other user benefits such as vehicle operating cost savings and some safety benefits, which are not considered in illustrating the costs and associated benefits, in the following examples.

Example: Estimating the Benefits of Proposed Improvements

Consider the design and construction of an additional lane on a 5-mi road to relieve congestion. Assume that the total cost of the proposed additional lane is \$6,000,000 and that the project is to be completed by year 2020. The present value of the project can be computed using

$$PV = \frac{FV}{(1+i)^n} \quad (1)$$

where
PV = Present value,
FV = Future value,
i = discount rate (3%), and
n = difference between the year of analysis and the year of costs and benefits being evaluated.

Assume that the average travel time savings per person after adding the lane amount to 25 s. Similarly, the total vehicle miles of travel (VMT) on this 5-mi facility are 10,000 vehicle miles per hour during morning and evening peak hours (total 4 h). Because there could be multiple passengers in a vehicle, consider the average occupancy rate of 1.20 passengers per vehicle to evaluate total passenger miles of travel (PMT) during each peak hour. The total PMT during morning and evening peak hours (2 h in the morning and 2 h in the evening, total 4 h) is 10,000 * 1.2 * 4 = 48,000. Table 1 summarizes the

Table 1. Total Present Costs and Benefits for the Proposed Design

Year	VMT during peak hours	Passenger miles travel	Travel time savings (s)	Total time savings for day (h)	Total cost savings per year	Present value of savings	Total costs	Present value of costs
2020	10,000	48,000	25.00	333.33	\$785,967	\$677,982	\$6,000,000	\$5,175,653
2021	10,150	48,720	24.27	328.48	\$774,521	\$648,649	\$10,000	\$8,375
2022	10,302	49,451	23.56	323.70	\$763,241	\$620,585	\$10,500	\$8,537
2023	10,457	50,193	22.88	318.98	\$752,126	\$593,735	\$11,025	\$8,703
2024	10,614	50,945	22.21	314.34	\$741,173	\$568,047	\$11,576	\$8,872
2025	10,773	51,710	21.57	309.76	\$730,379	\$543,470	\$12,155	\$9,045
2026	10,934	52,485	20.94	305.25	\$719,742	\$519,957	\$12,763	\$9,220
2027	11,098	53,273	20.33	300.80	\$709,261	\$497,461	\$13,401	\$9,399
2028	11,265	54,072	19.74	296.42	\$698,932	\$475,938	\$14,071	\$9,582
2029	11,434	54,883	19.16	292.10	\$688,753	\$455,347	\$14,775	\$9,768
2030	11,605	55,706	18.60	287.85	\$678,723	\$435,646	\$15,513	\$9,957
2031	11,779	56,542	18.06	283.66	\$668,838	\$416,798	\$16,289	\$10,151
2032	11,956	57,390	17.53	279.53	\$659,098	\$398,765	\$17,103	\$10,348
2033	12,136	58,251	17.02	275.46	\$649,499	\$381,512	\$17,959	\$10,549
2034	12,318	59,124	16.53	271.45	\$640,041	\$365,006	\$18,856	\$10,754
2035	12,502	60,011	16.05	267.49	\$630,720	\$349,214	\$19,799	\$10,962
2036	12,690	60,911	15.58	263.60	\$621,534	\$334,105	\$20,789	\$11,175
2037	12,880	61,825	15.13	259.76	\$612,483	\$319,650	\$21,829	\$11,392
2038	13,073	62,752	14.68	255.97	\$603,563	\$305,821	\$22,920	\$11,613
2039	13,270	63,694	14.26	252.25	\$594,774	\$292,589	\$24,066	\$11,839
2040	13,469	64,649	13.84	248.57	\$586,112	\$279,930	\$25,270	\$12,069
2041	13,671	65,619	13.44	244.95	\$577,576	\$267,819	\$26,533	\$12,303
2042	13,876	66,603	13.05	241.39	\$569,165	\$256,232	\$27,860	\$12,542
2043	14,084	67,602	12.67	237.87	\$560,876	\$245,146	\$29,253	\$12,786
2044	14,295	68,616	12.30	234.41	\$552,708	\$234,540	\$30,715	\$13,034
2045	14,509	69,645	11.94	230.99	\$544,659	\$224,392	\$32,251	\$13,287
Total						\$10,708,338	Total	\$5,441,915
Overall benefits								\$5,266,423

total cost savings for the proposed design considering the analysis timeframe of 25 years. Many values considered in the table can be obtained from the regional travel demand forecasting model outputs.

In the table, column 1 shows the time frame year, column 2 shows the VMT with an incremental rate of 1.5% each year, and column 3 shows the PMT for each year. With an increase in VMT, the travel time savings could decrease. Column 4 shows the travel time savings for each year with a 3% decrease in travel time savings, each year, from the previous year. Column 5 shows total travel time savings each day in hours and column 6 shows the total cost savings for each year. The cost savings are converted to 2015 dollars and are presented in column 7. Column 8 shows the cost incurred for implementation of the proposed design and other maintenance costs, each year, and column 9 shows the costs incurred in 2015 dollars. Overall, the proposed project will result in savings equal to $10,708,338 just because of reduced travel times. The savings could be even higher by considering reduced vehicle maintenance costs and improved safety. Considering a timeframe of 25 years, the overall benefit of the proposed design is $5,266,423.

Like the above example, a real-world scenario is considered to illustrate the benefits of proposed improvements. For this purpose, widening of a 10.3-mi section from four lanes to six lanes on I-40 South of Trinity Road, Raleigh is considered. The travel times decreased by 3.7 min and 5.5 min during morning peak hour and evening peak hour, after widening the road. With $6.46 per hour travel time saving value, the traffic volume after completion of road widening is considered to evaluate the cost savings from widening the road.

Table 2 shows the overall cost savings during peak hours on widening the I-40 section. The cost savings are estimated as $3,059 and $4,974 for morning peak hour and evening peak hour, respectively. Considering an overall of 4 h of peak volume on a given day (2 h in the morning and 2 h in the evening), the total cost savings in a year just during peak hours, because of widening the 10.3-mi section, are estimated as $5.86 million.

Example: Evaluating Transportation Alternatives

For illustration, two hypothetical alternative designs (Alternative A and Alternative B) for a road alignment between two points are considered. The cost of Alternative A is $80 million and the cost of Alternative B is $100 million. After the construction of the alignment in year 2020, assume the average VMT per day on the proposed road is 120,000. Table 3 shows the comparison of benefits and costs between the two alternatives. Many values considered in the table (from column 2 to column 6) can be obtained from the regional travel demand forecasting model outputs.

In Table 3, column 1 shows the year of analysis with a timeframe of 25 years; columns 2 and 3 show the VMT for both alternatives (which are the same) each year with a 1.5% increase in VMT each year; columns 4 and 5 show the PMT for both alternatives considering an average vehicle occupancy rate of 1.20; column 6 shows the travel time savings for Alternative B compared with Alternative A; column 7 shows the total time saving each day in hours; column 8 shows the total cost benefits each year from 2020 to 2045; column 9 shows the cost benefits as 2015 dollars; and, columns 10 to 13 show the overall costs including maintenance, each year, along with their respective 2015 costs.

The benefit–cost (B/C) ratio will help agencies evaluate transportation alternatives easily. The benefit–cost ratio is calculated as the ratio of total present value of the benefits between alternatives (sum of discounted benefits) to the discounted value of incremental costs (difference between the present values of cost of both the alternatives). If the B/C ratio is greater than 1, then the alternative considered is beneficial compared with the other alternative. From Table 3, the overall cost benefits from Alternative B over Alternative A are $25,453,070. The cost difference between the alternatives is $19,919,799 (Alternative B total cost – Alternative A total cost). The B/C ratio is 1.28, which is to say that Alternative B is more beneficial when compared with Alternative A, with a return on investments of $1.28 for

Table 2. Cost Benefits from Widening of I-40

Widening from four-lane to six-lane: I-40, South of Trinity Road

	Traffic volumes		Travel times (min)			
Time of day	Before	After	Before	After	Travel time savings (min)	Total cost savings
*a.m. peak	6,700	6,400	13.0	9.3	3.7	$3,059
*p.m. peak	6,800	7,000	15.5	10.0	5.5	$4,974

Note: *a.m. peak values are for westbound traffic and p.m. peak values are for eastbound.

Table 3. Comparison of Costs and Benefits between Two Alternatives

Year	VMT per day		PMT		Diff. in travel times (s)	Time savings per day (h)	Total cost savings per year	Present value of savings	Total costs		Present value of costs	
	Alt. A	Alt. B	Alt. A	Alt. B					Alt. A	Alt. B	Alt. A	Alt. B
2020	120,000	120,000	144,000	144,000	12.00	480.00	$1,131,792	$976,294	$80,000,000	$100,000,000	$69,008,703	$86,260,878
2021	121,800	121,800	146,160	146,160	12.18	494.51	$1,166,000	$976,507	$500,000	$600,000	$418,742	$502,491
2022	123,627	123,627	148,352	148,352	12.36	509.45	$1,201,243	$976,720	$525,000	$630,000	$426,873	$512,248
2023	125,481	125,481	150,578	150,578	12.55	524.85	$1,237,550	$976,934	$551,250	$661,500	$435,162	$522,194
2024	127,364	127,364	152,836	152,836	12.74	540.72	$1,274,955	$977,147	$578,813	$694,575	$443,612	$532,334
2025	129,274	129,274	155,129	155,129	12.93	557.06	$1,313,491	$977,361	$607,753	$729,304	$452,225	$542,670
2026	131,213	131,213	157,456	157,456	13.12	573.90	$1,353,191	$977,574	$638,141	$765,769	$461,006	$553,208
2027	133,181	133,181	159,818	159,818	13.32	591.24	$1,394,091	$977,788	$670,048	$804,057	$469,958	$563,950
2028	135,179	135,179	162,215	162,215	13.52	609.11	$1,436,228	$978,001	$703,550	$844,260	$479,083	$574,900
2029	137,207	137,207	164,648	164,648	13.72	627.52	$1,479,638	$978,215	$738,728	$886,473	$488,386	$586,063
2030	139,265	139,265	167,118	167,118	13.93	646.49	$1,524,360	$978,428	$775,664	$930,797	$497,869	$597,443
2031	141,354	141,354	169,625	169,625	14.14	666.03	$1,570,433	$978,642	$814,447	$977,337	$507,537	$609,044
2032	143,474	143,474	172,169	172,169	14.35	686.16	$1,617,900	$978,856	$855,170	$1,026,204	$517,392	$620,870
2033	145,626	145,626	174,752	174,752	14.56	706.90	$1,666,801	$979,070	$897,928	$1,077,514	$527,438	$632,926
2034	147,811	147,811	177,373	177,373	14.78	728.27	$1,717,180	$979,284	$942,825	$1,131,389	$537,680	$645,216
2035	150,028	150,028	180,033	180,033	15.00	750.28	$1,769,082	$979,498	$989,966	$1,187,959	$548,120	$657,744
2036	152,278	152,278	182,734	182,734	15.23	772.96	$1,822,552	$979,712	$1,039,464	$1,247,357	$558,763	$670,516
2037	154,562	154,562	185,475	185,475	15.46	796.32	$1,877,639	$979,926	$1,091,437	$1,309,725	$569,613	$683,536
2038	156,881	156,881	188,257	188,257	15.69	820.39	$1,934,390	$980,140	$1,146,009	$1,375,211	$580,673	$696,808
2039	159,234	159,234	191,081	191,081	15.92	845.18	$1,992,857	$980,354	$1,203,310	$1,443,972	$591,949	$710,338
2040	161,623	161,623	193,947	193,947	16.16	870.73	$2,053,092	$980,568	$1,263,475	$1,516,170	$603,443	$724,131
2041	164,047	164,047	196,856	196,856	16.40	897.05	$2,115,146	$980,782	$1,326,649	$1,591,979	$615,160	$738,192
2042	166,508	166,508	199,809	199,809	16.65	924.16	$2,179,077	$980,996	$1,392,981	$1,671,578	$627,105	$752,526
2043	169,005	169,005	202,806	202,806	16.90	952.09	$2,244,939	$981,211	$1,462,630	$1,755,156	$639,282	$767,138
2044	171,540	171,540	205,848	205,848	17.15	980.87	$2,312,792	$981,425	$1,535,762	$1,842,914	$651,695	$782,034
2045	174,113	174,113	208,936	208,936	17.41	1010.52	$2,382,697	$981,639	$1,612,550	$1,935,060	$664,349	$797,219
Total								$25,453,070	Total		$82,321,818	$102,236,617
Benefit–cost ratio											1.28	

Note: Alt. = Alternative.

every $1 spent. The procedure is repeated until all alternatives are considered to evaluate the most economically feasible alternative.

Reliability by Time-of-the-Day: Illustration

The reliability of a link or segment could vary based on the time-of-the-day. The possibility of capturing extensive, continuous, and dynamic travel time data from private sources such as HERE, INRIX and Tom Tom opened many pragmatic avenues to predict reliability at link-level by time-of-the-day (*22*, *23*). BT was observed to be strongly correlated to most travel time performance measures and may be suitable for assessing reliability (*22*, *23*). As sum of deviations is not the same as deviations of sums, data/measures should be carefully integrated when conducting path-level assessments (not used in the illustrations discussed in this paper).

Data from INRIX, for the year 2011, was used to illustrate the reliability of links by time-of-the-day and day-of-the-week (representative of general traffic conditions, with no construction activity within the vicinity of study segments). Travel time performance measures such as average travel time and 95th percentile travel time were computed by time-of-the-day and day-of-the-week for road links in the Charlotte metropolitan area. For illustration purposes, a 2.1-mi segment on I-77 northbound at the Harris Oak Boulevard/Reames Road/Exit 18 intersection and a 3.6-mi segment on I-77 northbound at the Gilead Road/Exit 23 intersection in the City of Charlotte area are considered.

Figure 1 shows the comparison of the average travel time and the 95th percentile travel time for the 2.1-mi segment on I-77 northbound at the Harris Oak Boulevard/Reames Road/Exit 18 intersection. The BT is the difference between the 95th percentile travel time and the average travel time at any given time-of-the-day. From Figure 1, it can be noticed that the 95th percentile travel times are always greater than the average travel times, irrespective of the time-of-the-day and day-of-the-week. This indicates that the segment is unreliable during all times of the day and days of the week.

Figure 2 shows the comparison of the average travel time and the 95th percentile travel time for the 3.6-mi segment on I-77 northbound at the Gilead Road/Exit 23 intersection. From Figure 2, it can be noticed that the 95th percentile travel time is always equal to the average travel time, except during evening peak hours from Monday through Saturday. However, the 95th percentile travel time is equal to the average travel time during any time-of-the-day on Sundays. This indicates that the segment is unreliable only during evening peak hours from Monday through Saturday and completely reliable on Sundays.

The reliability (total BT) of the link during a day-of-the-week can be evaluated by computing the area between the average travel time and 95th percentile travel time curves in Figures 1 and 2. For illustration, Figure 3 shows the area between the average travel time and the 95th percentile travel time for the 2.1-mi segment on I-77 northbound at the Harris Oak Boulevard/Reames Road/Exit 18 intersection on Mondays.

The BT between the curves during a peak hour (8:00 to 9:00 a.m.) is 15 s. Multiplying the BT with traffic volume between 8:00 and 9:00 a.m. (assume 6,000 vehicles) will give the total BT (1,500 min) during the considered peak period. Multiplying the total BT with the generalized value of reliability ($ 0.45 per minute) will help estimate the total value of unreliability ($600) during the considered peak hour. In case of availability of hourly traffic volumes for the entire day, the area in plots like Figure 3 with hourly traffic volumes on the *x* axis and BTs on the *y* axis will directly give the total BT for the time period considered. Multiplying the total BT with the generalized value of reliability will help monetize the value of reliability for the road segment considered by time-of-the-day and day-of-the-week.

Value of BT: Illustration

As discussed earlier, reliability accounts for consistency in travel times. Travelers account for the variability in the travel time and plan for BT along with the average travel time to their destination. The unreliability increases as the BT increases. The cost associated with the planned BT varies when compared with the travel time savings. For example, a traveler may plan for a BT of 15 min to his/her 25-min trip. The planned additional 15 min is because of the uncertainty in the travel time between the origin and destination (to ensure that he/she reaches a destination on time). As the uncertainty in the travel times on the road links between the origin and destination decreases, the planned BT will also decrease. The value of the planned BT is different when compared with the travel time savings for the 25-min trip. It also varies by time-of-the-day and day-of-the-week. The following example illustrates the use of the value of BT in estimating benefits.

Example: Estimating the Benefits from a Reliable Route

Consider a 10-mi road link with average travel times as presented in column 2 of Table 4. Column 3 is BT for each trip during the given hour. Similarly, the traffic volumes on the given road for each hour between 07:00 a.m. and 10:00 p.m. are shown in column 4. Column 5 shows the total passengers traveled with 1.2 as the

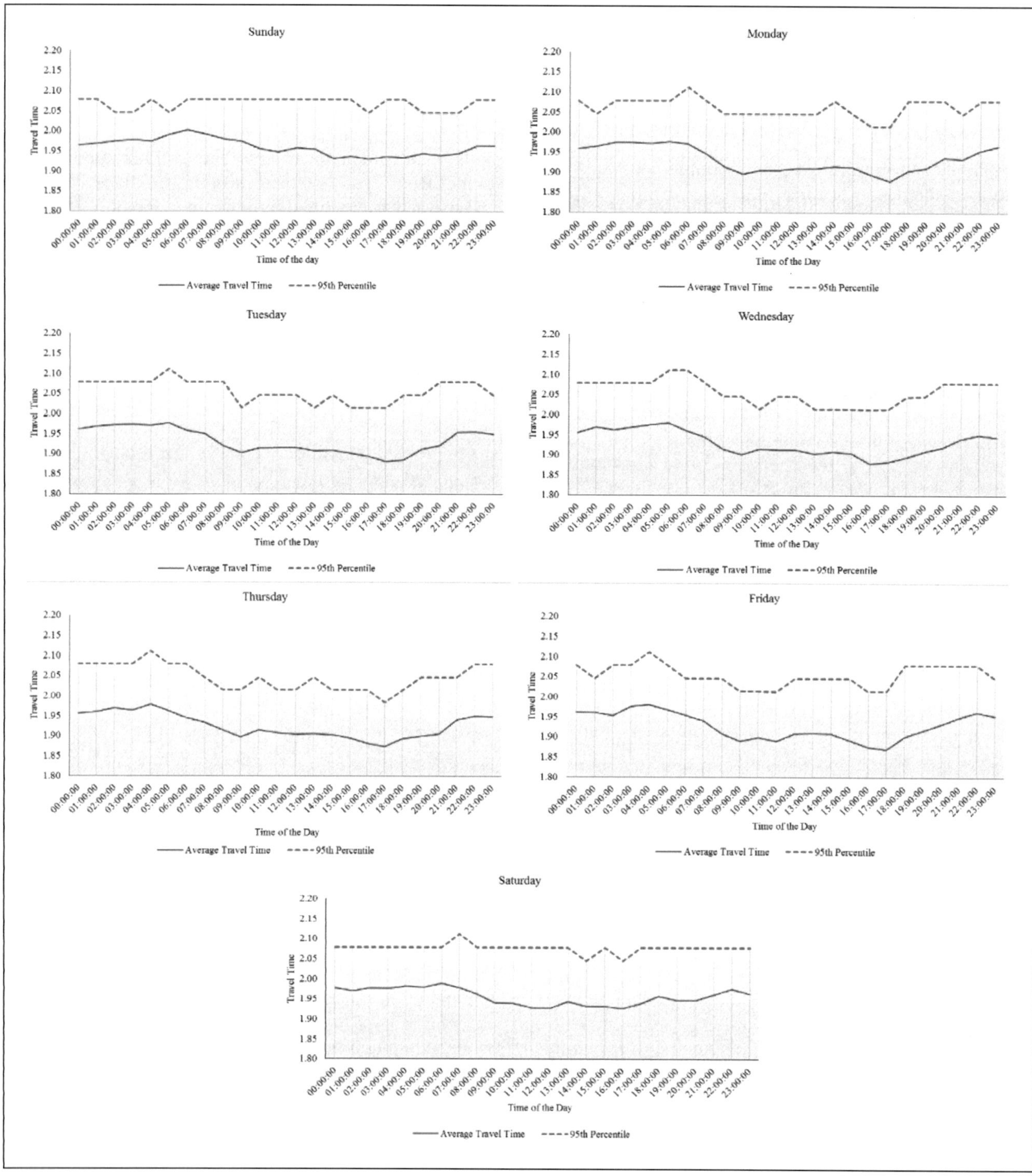

Figure 1. Segment that is unreliable during all times of the day and days of the week.

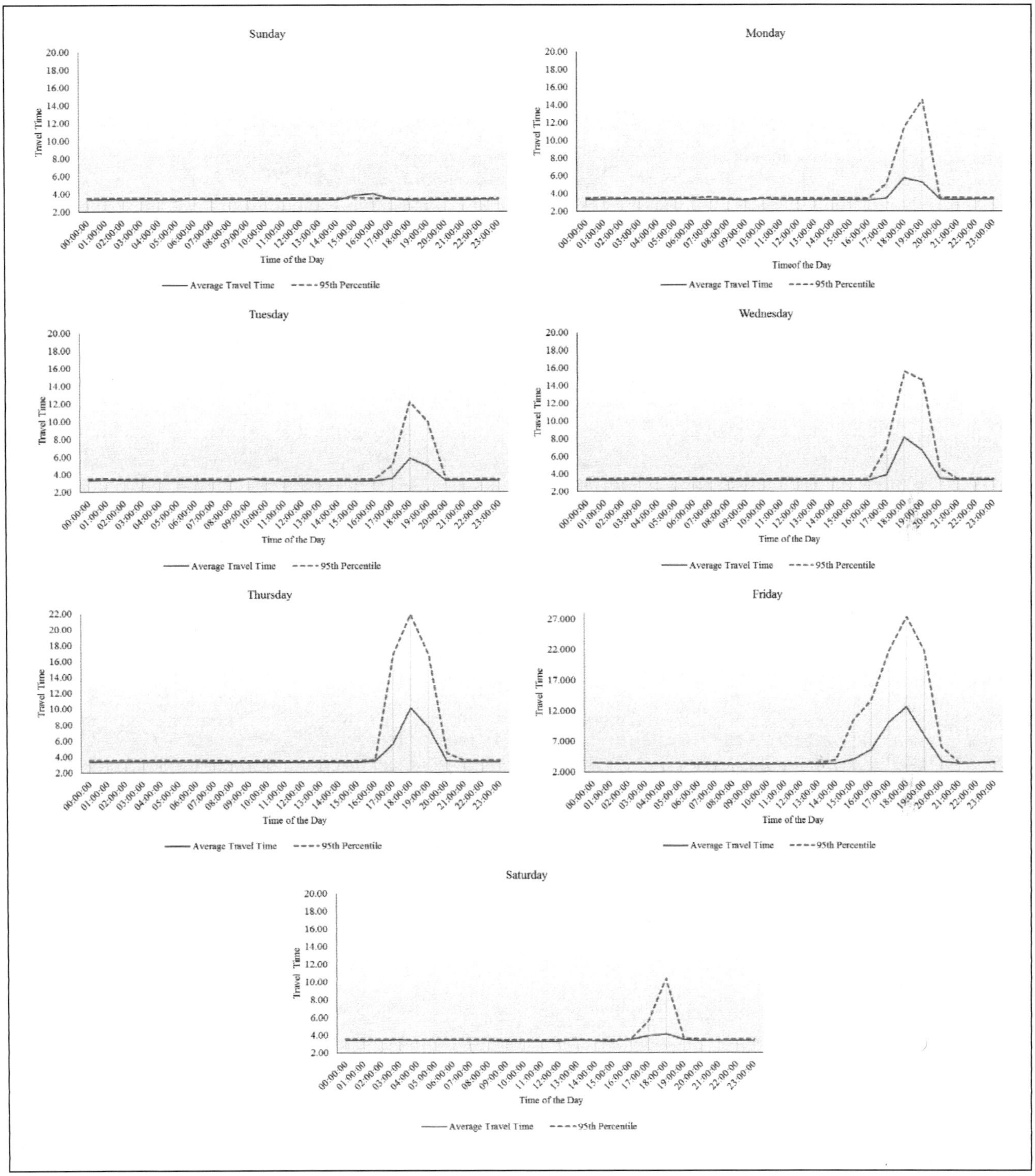

Figure 2. Segment that is unreliable during evening peak hours.

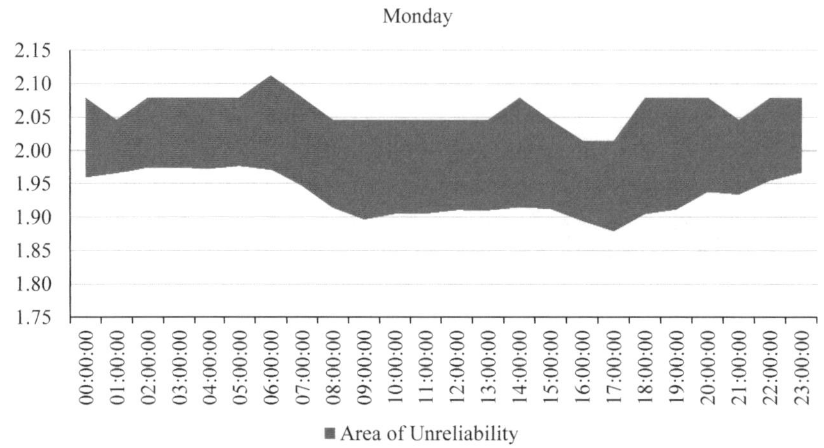

Figure 3. Area between average and 95th percentile travel time curves.

Table 4. Estimation of Cost Benefits from a Reliable Route by Time-of-the-Day

Time-of-the-day	Average travel time (min)	Buffer time (min)	Traffic volume	Total passengers	Value of travel time per year	Value of BT per year
07:00 to 08:00 a.m.	25	10	4,000	4,800	$22,338,000	$7,884,000
08:00 to 09:00 a.m.	26	9	4,200	5,040	$24,393,096	$7,450,380
09:00 to 10:00 a.m.	23	10	3,800	4,560	$19,523,412	$7,489,800
10:00 to 11:00 a.m.	22	3	3,500	4,200	$17,200,260	$2,069,550
11:00 a.m. to 12:00 noon	21	4	3,400	4,080	$15,949,332	$2,680,560
12:00 noon to 01:00 p.m.	20	3	3,200	3,840	$14,296,320	$1,892,160
01:00 to 02:00 p.m.	20	3	3,000	3,600	$13,402,800	$1,773,900
02:00 to 03:00 p.m.	20	3	2,600	3,120	$11,615,760	$1,537,380
03:00 to 04:00 p.m.	21	3	3,200	3,840	$15,011,136	$1,892,160
04:00 to 05:00 p.m.	23	5	3,700	4,440	$19,009,638	$3,646,350
05:00 to 06:00 p.m.	26	10	4,400	5,280	$25,554,672	$8,672,400
06:00 to 07:00 p.m.	27	10	4,200	5,040	$25,331,292	$8,278,200
07:00 to 08:00 p.m.	24	9	3,900	4,680	$20,908,368	$6,918,210
08:00 to 09:00 p.m.	23	6	3,450	4,140	$17,725,203	$4,079,970
09:00 to 10:00 p.m.	20	3	2,500	3,000	$11,169,000	$1,478,250
Total					$273,428,289	$67,743,270

average vehicle occupancy rate. Column 6 shows the value of travel time, and column 7 shows the value of BT. For this transportation project or alternative, the total value of travel time is $273,428,289 per year, whereas the total value of BT is $67,743,270 per year.

Like the above example, a real-world scenario is considered to illustrate the value of reliability. For this purpose, widening of a 10.3-mi section from four lanes to six lanes on I-40 South of Trinity Road, Raleigh is considered. After widening the road, the difference between the maximum travel time (95th percentile travel time is not available for illustration) and the average travel time decreased from 9 min to 2 min during the morning peak hour. Similarly, the difference between the maximum travel time and the average travel time reduced from 10.5 min to 2.7 min during the evening peak hour.

The generalized value of BT is $ 0.45 per minute. The traffic volumes after completion of road widening are considered to evaluate the value of improved reliability from road widening. Table 5 shows the overall value of the project because of improved reliability during peak hours, on widening the I-40 section. A quick comparison with benefits created by a decrease in travel time for the same section indicates greater benefits achieved because of reduction in BT along this I-40 section.

Table 5. Cost Benefits of Improved Reliability from Widening of I-40

Widening from four-lane to six-lane: I-40, South of Trinity Road						
	Traffic volumes		Buffer (min)			
Time-of-the-day	Before	After	Before	After	Reduction in BT (min)	Value of reliability
*a.m. peak	6,700	6,400	9.0	2.0	7.0	$20,160
*p.m. peak	6,800	7,000	10.5	2.7	7.8	$24,570

Note: *a.m. peak values are for westbound and p.m. peak values are for eastbound.

Conclusion

The average travel time and BT (difference between the 95th percentile travel time and the average travel time) are the performance measures that are considered for illustration to evaluate transportation projects or alternatives. This hourly data required for evaluation could be easily extracted from private data sources or other means. The monetary values for average travel time and BT are $ 0.51 and $ 0.45 per minute, respectively.

The illustration of the use of monetary values indicates that benefits achieved through reduction in BT are substantially higher compared with benefits from decrease in travel time for the same section. The illustration of use of monetary value of BT helps agencies in prioritizing transportation projects or alternatives. The processes help the agencies efficiently and effectively use the limited available resources.

The effect and impact of transportation projects or alternatives are monetized through illustrations to evaluate and make improved decisions. Both improved travel time and reliability on roads yield significant monetary benefits. However, greater benefits can be achieved through improved reliability compared with benefits from decrease in travel time for a given section of road.

The outcomes from this research to monetize reliability for evaluating the impact of transportation alternatives is completely based on the surveys obtained from passenger car users (*20*). Therefore, the illustrations presented in this paper are only applicable for evaluating the impact based on the value of travel time reliability for passenger cars. The findings are not applicable for other modes of travel or vehicle types (such as trucks).

There are many ways to address and reduce recurring and nonrecurring congestion. Only long-term solutions (such as construction projects) were considered for illustration purposes in this research. However, the illustrative approaches could be adopted or adapted to conduct assessments of other solutions and strategies (for example, freeway management patrols, incident management, etc.).

The illustrations presented in this paper are based on generalized values for the state of North Carolina. It is likely that tolerance levels and the value of reliability could vary based on the trip lengths (time or distance). Incorporating such variations in assessing the value of reliability merits an investigation.

Unreliable travel times for truckers could lead to late shipments and disruption of on-time delivery, resulting in loss of competitive edge over other shippers. Therefore, the value of travel time reliability for trucks could be very high compared with passenger car travel time reliability. Effective methods for evaluating travel time reliability for freight and monetizing reliability for trucks (freight transportation) also merit an investigation.

Acknowledgments

The authors acknowledge the NCDOT for providing financial support for this project. Special thanks are extended to J. Kevin Lacy, Kelly Wells, Dwayne Alligood, Cheryl Evans, Terry Hopkins, Ehren Meister, David Wasserman, A. Whitley IV, Terry Arellano, Joseph Geigle, Meredith McDiarmid, Alpesh Patel, Anthony Wyatt and Ernest Morrison of NCDOT for providing excellent support, guidance, and valuable inputs for successful completion of this project. In addition, data collection efforts by transportation engineering graduate students of the Department of Civil & Environmental Engineering at The University of North Carolina at Charlotte (UNC Charlotte) is also appreciated and recognized.

Author Contributions

The authors confirm contribution to each section of the paper. All the authors reviewed the results and approved the final version of the manuscript.

References

1. Hallenbeck, M. E., J. M. Ishimaru, and J. Nee. *Measurement of Recurring Versus Non-Recurring Congestion: Technical Report.* Washington State Transportation Center, Seattle, 2003.
2. *Traffic Congestion and Reliability: Trends and Advanced Strategies for Congestion Mitigation.* Federal Highway Administration, 2005. http://www.ops.fhwa.dot.gov/congestion_report/executive_summary.htm. Accessed July 15, 2015.

3. Pisarski, A. *NCHRP Report 550: Commuting in America III*. Transportation Research Board of the National Academies, Washington, D.C., 2007.
4. Pulugurtha, S. S., and N. Pasupuleti. Assessment of Link Reliability as a Function of Congestion Components. *Journal of Transportation Engineering*, Vol. 136, No. 10, 2010, pp. 903–913.
5. Ebeling, C. E. *Introduction to Reliability and Maintainability Engineering*. McGraw-Hill, New York, 1997.
6. *Travel Time Reliability: Making It There on Time, All the Time*. Federal Highway Administration, 2014. www.ops.fhwa.dot.gov/publications/tt_reliability/TTR_Report.htm. Last Accessed July 14, 2014.
7. Lomax, T., D. Schrank, S. Turner, and R. Margiotta. *Selecting Travel Reliability Measures*. Texas Transportation Institute Monograph, College Station, 2003. http://d2dtl5nnlpfr0r.cloudfront.net/tti.tamu.edu/documents/TTI-2003-3.pdf. Accessed February 15, 2015.
8. Moses, L. N., and H. F. Williamson. Value of Time, Choice of Mode, and the Subsidy Issue in Urban Transportation. *The Journal of Political Economy*, Vol. 71, No. 3, 1963, pp. 247–264.
9. Becker, G. S. A Theory of the Allocation of Time. *The Economic Journal*, Vol. 75, No. 299, 1965, pp. 493–517.
10. Lisco, T. E. Value of Commuters Travel Time: A Study in Urban Transportation. *Highway Research Record*, No. 245, 1968, p. 36.
11. Reichman, S. Subjective Time Savings in Interurban Travel: An Empirical Study. *Highway Research Record*, No. 446, 1973, pp. 21–27.
12. Cherlow, J. R. Measuring Values of Travel Time Savings. *Journal of Consumer Research*, Vol. 7, No. 4, 1981, pp. 360–371.
13. Litman, T. *Transportation Cost Analysis for Sustainability*. Victoria Transportation Policy Institute, 1999.
14. Kittelson, W., and M. Vandehey. SHRP2: *Incorporation of Travel Time Reliability into the HCM*. Transportation Research Board of National Academies, Washington, D.C., 2013.
15. Abdel-Aty, M. A., R. Kitamura, and P. P. Jovanis. Using Stated Preference Data for Studying the Effect of Advanced Traffic Information on Drivers' Route Choice. *Transportation Research Part C: Emerging Technologies*, Vol. 5, No. 1, 1997, pp. 39–50.
16. Lam, T. C., and K. A. Small. The Value of Time and Reliability: Measurement from a Value Pricing Experiment. *Transportation Research Part E: Logistics and Transportation Review*, Vol. 37, No. 2–3, 2001, pp. 231–251.
17. Nam, D., D. Park, and A. Khamkongkhun. Estimation of Value of Travel Time Reliability. *Journal of Advanced Transportation*, Vol. 39 No. 1, 2005, pp. 39–61.
18. Tilahun, N. Y., D. M. Levinson, and K. J. Krizek. Trails, Lanes, or Traffic: Valuing Bicycle Facilities with an Adaptive Stated Preference Survey. *Transportation Research Part A: Policy and Practice*, Vol. 41, No. 4, 2007, pp. 287–301.
19. He, X., X. H. Liu, and X. J. Cao. Estimating Value of Travel Time and Value of Reliability Using Dynamic Toll Data. Presented at 91st Annual Meeting of the Transportation Research Board, Washington, D.C., 2012.
20. Pulugurtha, S. S., P. Penmetsa, and V. R. Duddu. *Monetizing Reliability to Evaluate the Impact of Transportation Alternatives*. Report # FHWA/NC/2015-07. North Carolina Department of Transportation, 2017.
21. Pulugurtha, S. S., V. R. Duddu, P. Penmetsa, V. M. Kukkapalli, S. Tagar, and S. R. Kodupuganti. Monetary Value of Travel Time Reliability and Thresholds based on Random Participants Perceptions. Presented at 97th Annual Meeting of the Transportation Research Board, Washington, D.C., 2018.
22. Pulugurtha, S. S., V. R. Duddu, R. Puvvala, V. R. Thokala, and M. I. Shah. *Commercial Remote Sensing & Spatial Information (CRS & SI) Technologies Program for Reliable Transportation Systems Planning: Volume 2 - Comparative Evaluation of Travel Time Related Performance Measures*. Report No. RITARS-12-H-UNCC-2. U.S. Department of Transportation, Washington, D.C., 2015.
23. Pulugurtha, S. S., V. R. Duddu, and V. R. Thokala. Travel-Time-Based Performance Measures: Examining Interrelationships and Recommendations for Analysis. Presented at 95th Annual Meeting of the Transportation Research Board, Washington, D.C., 2016.

The Standing Committee on Transportation Programming and Investment Decision-Making (ADA50) peer-reviewed this paper (18-05371).

Article

TRR
JOURNAL OF THE TRANSPORTATION RESEARCH BOARD

Transportation Research Record
2018, Vol. 2672(51) 99–108

DOI: 10.1177/0361198118782801
journals.sagepub.com/home/trr
SAGE

Programming Flexible Congestion Mitigation and Air Quality Program Funds: Best Practices for State DOTs

Gregory Rowangould[1], Razieh Nadafianshahamabadi[1], and Amir Poorfakhraei[1]

Abstract

State departments of transportation (DOTs) face unique challenges in programming funds for the federal Congestion Mitigation and Air Quality Improvement (CMAQ) program. This is particularly true as more DOTs receive larger shares of so-called "flexible" CMAQ funding. Unlike other CMAQ funds, flexible funds can be spent outside of non-attainment and maintenance areas, which often means in more rural areas where there is no metropolitan planning organization (MPO) to program the funding. Flexible funds may also be spent on a wider range of project types. DOTs in these situations have many options for programming CMAQ funds but there has been little guidance on best practices for doing this. Prior studies have focused exclusively on best practice for MPOs. This study provides a comprehensive survey of current CMAQ programming practices among DOTs and MPOs and then highlights practices best suited to the particular needs of DOTs.

This paper is derived from a more comprehensive study conducted by the authors for the New Mexico Department of Transportation (NMDOT). The aim of the study was to assist NMDOT in developing a robust and efficient framework for managing its federal Congestion Mitigation and Air Quality Improvement (CMAQ) program funding by surveying the current state of practice in the solicitation, evaluation, and selection of transportation projects that are eligible for CMAQ program funding, and to recommend best practices for completing these tasks. We believe many of the findings and recommendations from our original study are likely applicable to challenges now facing many state DOTs and are therefore of broad appeal and use. This paper summarizes the most broadly applicable and important findings from the work commissioned by NMDOT.

A particular focus of our study was on the practice and challenges in states that receive so-called "flexible" CMAQ funds. These are states that generally have relatively modest transportation-related air quality problems and therefore are given greater flexibility in how CMAQ funds can be used. One outcome of greater flexibility is that less funds are required to be spent in metropolitan areas, providing state DOTs with greater flexibility in how and where these funds can be used.

While several prior studies have evaluated the CMAQ program, their focus and aims have been somewhat different than our study. The Transportation Research Board (*1*) and Federal Highway Administration (FHWA) (*2*) conducted case studies that broadly evaluated the practices used to program CMAQ funds by 12 metropolitan planning organizations (MPOs) in total. Several other studies have evaluated the cost effectiveness of different types of projects commonly funded by the CMAQ program at reducing vehicle emissions (*1*, *3*, *4*) or the effectiveness of the program as a whole (*5*). One limitation of these past studies is that they focused almost exclusively on urban areas facing significant air quality challenges and the CMAQ programming practices of MPOs. While many of the findings of these studies may be relevant for state DOTs, they do not focus on the unique challenges that many DOTs face.

There are several significant differences between MPOs and DOTs that cause different challenges in programming CMAQ funds. The main difference is that a DOT generally acts as the regional planning agency for areas outside of MPOs. These are for the most part rural areas that have very different transportation systems and challenges than the urban areas overseen by MPOs. In these areas, air quality and congestion problems are more likely to be limited to specific highway corridors rather than large urban areas. Therefore, CMAQ project cost effectiveness guidelines based on experience in urban areas may not be that relevant for rural communities and small towns. Furthermore, the technical resources of rural communities and small towns are

[1]Department of Civil Engineering, University of New Mexico, Albuquerque, NM

Corresponding Author:
Address correspondence to Gregory Rowangould: rowangould@unm.edu

likely much less than those of larger cities and MPOs. The lack of technical resources may make it more difficult to develop CMAQ project funding proposals, which in many cases require more quantitative analysis than typical funding proposals, for example, a quantitative estimate of emissions reductions. While DOT staff may have technical resources to aid local communities in developing proposals, the areas overseen by DOTs can be very large, causing resources to be spread very thin. Lastly, DOTs with larger shares of flexible funds need to determine how to allocate funds between urban areas planned by MPOs and other areas planned by the DOT and rural communities and small towns.

Overview of the CMAQ Program

The CMAQ program was established by the 1991 Intermodal Surface Transportation Efficiency Act (ISTEA) with the objective of funding transportation programs and projects that would contribute to attainment of the National Ambient Air Quality Standards (NAAQS). It has been re-authorized in each subsequent federal transportation bill. A wide range of transportation projects and programs linked to reducing vehicle emissions in non-attainment and maintenance areas are eligible for CMAQ funds. Each state receives a minimum CMAQ apportionment regardless of whether it has now or has ever had any non-attainment or maintenance areas. The minimum apportionment is a legacy carried forward from SAFETEA-LU, and prior transportation bills, which guaranteed states a minimum of one-half a percent of total CMAQ program funding regardless of whether they had any air quality problems. In states with non-attainment and maintenance areas, a portion of the CMAQ apportionment must be spent in non-attainment and maintenance areas. The exact amount depends on the size of the population in non-attainment and maintenance areas, the air pollutant(s) involved, and the severity of air quality problems (*6*). In states such as California that have large populations in non-attainment and maintenance areas, all CMAQ funds must be spent in non-attainment and maintenance areas. However, in states with less widespread air quality problems, a portion of the CMAQ funds (up to 100% in some cases) may be used outside of non-attainment and maintenance areas. These funds are commonly referred to as "flexible" CMAQ funds. Flexible CMAQ funds can also be used for activities that are eligible for funds from the Surface Transportation Block Grant (STBG) program in addition to the CMAQ program. This flexibility allows CMAQ funds to be spent on a much wider range of transportation projects, including those that have no air quality benefit, since most types of surface transportation projects are eligible for STBG funds.

Despite the many requirements and restrictions for using federal-aid highway funds, and CMAQ funds in particular, states are afforded nearly complete discretion in how to evaluate the emission reduction benefits of CMAQ projects. The FHWA does expect states to estimate emission reductions for all CMAQ projects quantitatively except when "it is not possible to accurately quantify emission benefits" and to weigh the cost effectiveness of emission reductions heavily in prioritizing projects for funding (*6*). However, there are no requirements for how a state calculates the emission reduction potential of a CMAQ project or how it weighs emission reductions against other project costs and benefits.

States and MPOs are also afforded wide discretion in the methods they use to solicit, rate, and ultimately select CMAQ projects. The FHWA's CMAQ Program Interim Guidance (*6*), however, does lay out the agency's expectations. These include that the project selection process be transparent, in writing, and publicly available; that agencies involved in rating proposed projects and committees involved in making final decisions are identified; and that the process and basis for rating proposed projects is clear. The guidance and the CMAQ legislation also encourage states and MPOs to coordinate with air quality agencies when making project selection decisions.

Methods

We sought information on how DOTs program CMAQ funds from three sources. First, we evaluated how DOTs and MPOs had spent CMAQ funds in the past five years, from 2011 to 2015, by evaluating the FHWA CMAQ Public Access System. The CMAQ Public Access System contains information about every CMAQ project completed since 1992, including project emission reduction estimates, costs, location, and description. We stratified our analysis of the database by DOT and MPO programmed projects that used flexible and non-flexible funds. This allowed us to compare differences in the types of projects funded by DOTs and MPOs and also in areas with and without significant air quality problems (i.e., in those in NAAQS non-attainment areas).

Projects that used flexible CMAQ funds are identified in the FHWA database as "Federal Surface Transportation Program/Congestion Mitigation and Air Quality (STP/CMAQ)" projects. Some of these projects are CMAQ-type projects and others are activities that are only eligible under the STP or STBG program. To complete our analysis, in most cases we were able to assign STP/CMAQ projects to one of the standard CMAQ project types or an additional category of "other" for activities that are ineligible for CMAQ funding based on a review of a project's title and description provided in the FHWA CMAQ database. In some cases, it was necessary to look up individual projects from the websites of MPOs or DOTs using their Transportation Improvement Program (TIP) or Statewide Transportation Improvement Program (STIP) identification numbers for more information.

We also searched DOT and MPO websites for information about how they manage their CMAQ programs. We searched every DOT website in states that receive some amount of flexible CMAQ funds. We also searched the websites of

MPOs that have sponsored CMAQ projects using flexible funds. There are over 400 MPOs and 56 of them received some flexible funds. Given the large number of MPOs and our focus on DOTs we selected just four that represent metropolitan areas of different sizes and that are in different areas of the country. The aim of the web search was to identify the range of methods currently used for soliciting, evaluating, and selecting CMAQ projects in areas that receive flexible CMAQ funds. We included the MPOs so that we could compare any differences in practice and to identify potentially useful MPO practices that could be used by DOTs.

Finally, we requested information about best practice in CMAQ program management from academics, consultants, and local, region, state, and federal agency staff who are experts in transportation and air quality. The request was made by sending out a brief e-mail to a list-serve maintained by the Transportation Research Board standing committee on Transportation and Air Quality (ADC20). The committee list-serve includes past and present members of the committee, researchers who have submitted research papers to the committee for consideration in prior Transportation Research Board annual meetings, and other researchers, consultants, and agency staff interested in transportation air quality research and policy.

Results

Funded CMAQ Projects

Most CMAQ projects are programmed by MPOs or occur in MPO areas. Between 2011 and 2015, 84% of CMAQ projects occurred in MPO areas that were also non-attainment or maintenance areas, while 7% occurred in MPO areas that were in attainment. Over this same time period, only 9.2% of CMAQ projects occurred outside of MPO areas, with more occurring in attainment areas (5.6%) than non-attainment areas (3.6%). There is little variation in the breakdown of projects from year to year, except for 2015 when the number of projects in MPO areas that were in attainment more than doubled from 92 in 2014 to 232 in 2015 and the number of projects in non-attainment and maintenance areas decreased from 1,407 in 2014 to 1,052 in 2015.

Overall, as shown in Figure 1, about 40% of all CMAQ projects and CMAQ funding are classified as congestion reduction and traffic flow improvements. Examples include intersection and traffic signal control improvements, incident management systems, high occupancy vehicle lanes (including tolled lanes), congestion and vehicle miles traveled (VMT) charging schemes, and a wide range of intelligent transportation system projects. The next most common types of project are bicycle and pedestrian projects, accounting for about 20% of CMAQ projects and just over 10% of CMAQ funding. Examples include construction of bicycle and pedestrian facilities that are not exclusively recreational, outreach programs for encouraging safe bicycle use, and funding for a state bicycle/pedestrian coordinator. The other major CMAQ project types are transit projects, accounting for close to 15% of CMAQ projects and 30% of CMAQ funding. Transit projects include construction and operating assistance for new transit projects, replacement of transit vehicles, and fare subsidies. Other project types account for less than 10% of CMAQ projects and CMAQ funding. Little

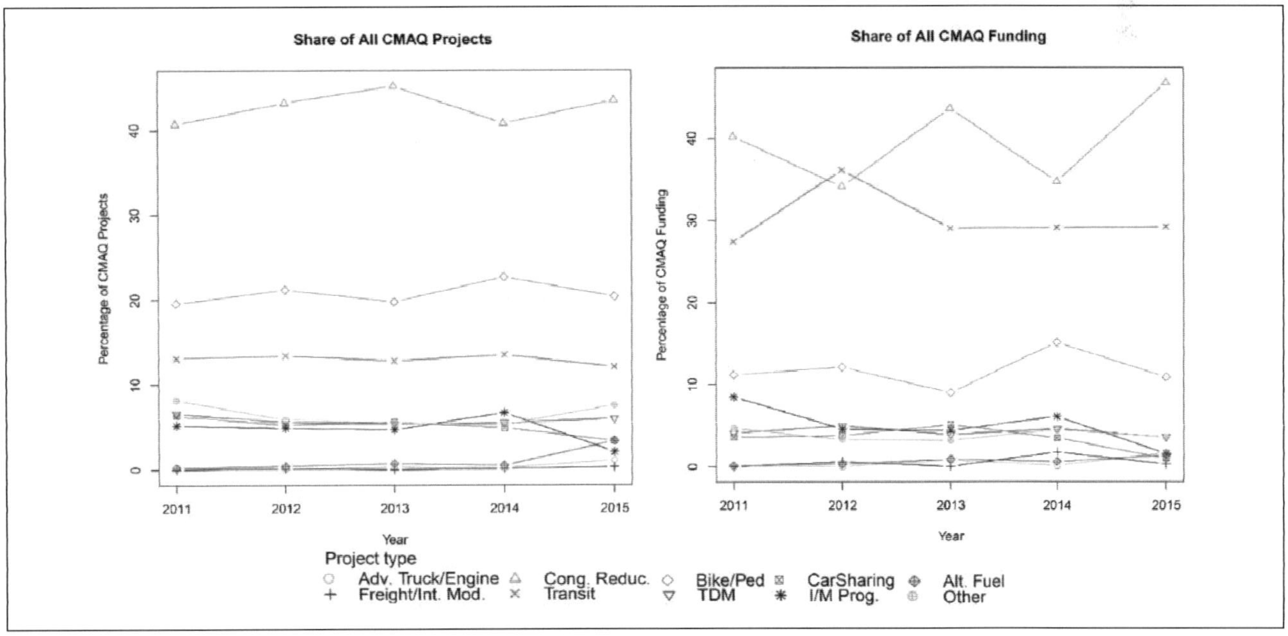

Figure 1. Share of projects funded by the CMAQ program by project type from 2011 to 2015.

to no funding has been used for freight and intermodal projects, diesel engine retrofits and advanced truck technologies, or alternative fuels and vehicles. There was very little year-to-year variation in the types of projects funded over the five-year period that we evaluated.

CMAQ Projects Funded by DOTs. As discussed above, only a small share of CMAQ projects occurred outside of MPO areas where DOTs programmed the projects. The projects programmed by DOTs were, as expected, different than those programmed by MPOs.

In attainment areas, 40% to 80% of CMAQ projects fell into the "other" category (Figure 2). This category represents projects that were entered into the FHWA CMAQ database as "STP/CMAQ" projects and which we further classified as being non-CMAQ eligible project types (i.e., STP or STBG projects). The share of CMAQ project funding spent on "other" projects was more variable and ranged between 5% and 70% over the five-year period.

In attainment areas, congestion reduction projects were also relatively common. They accounted for approximately 10% to 30% of projects and 10% to 50% of funding. The frequency of projects in other CMAQ categories was relatively low. The amount of funding for projects in other CMAQ categories was also relatively small, but more variable. Just under 10% of funding per year was spent on bicycle and pedestrian projects. Transit, freight, and intermodal projects received relatively large shares of funding in some years and nearly nothing in others.

CMAQ projects that occurred outside of MPO areas that were also in non-attainment or maintenance areas were different than those that occurred in attainment areas (Figure 2).

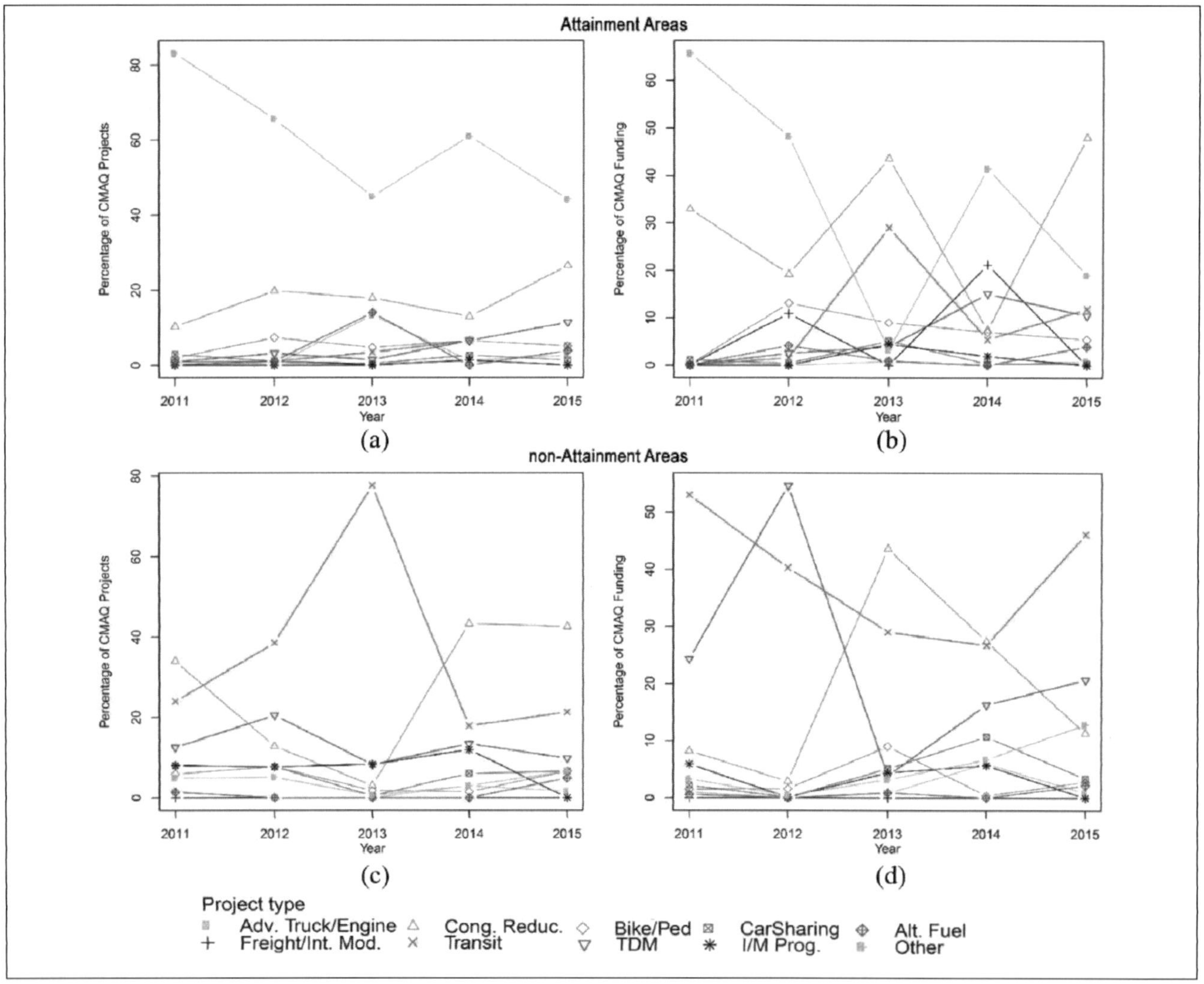

Figure 2. Share of state DOT projects funded by the CMAQ program by project type from 2011 to 2015 in attainment areas (*a* and *b*) and non-attainment areas (*c* and *d*).

Transit and congestion management projects were most common and accounted for the largest shares of funding. Travel demand management (TDM) projects were also more common, accounting for 10% to 20% of projects and between 5% and 55% of funding. Inspection and maintenance programs were also common, though they did not account for a large share of funding.

The priorities of DOTs, which accounted for a very small share of CMAQ projects, were somewhat different than MPOs. DOTs spent a greater share of flexible CMAQ funds on non-CMAQ eligible projects (i.e., STP projects) than MPOs did. Transit, pedestrian, and bicycle projects were also less common when a DOT was the sponsor, while TDM and inspection and maintenance programs were relatively more common. This difference between DOTs and MPOs is likely due to DOT projects taking place in more rural areas or on state highways where bicycle, pedestrian, and transit projects would be less effective or practical. Overall, there were no notable changes in the frequency of project types or funding over the analysis period.

CMAQ Project Selection Practices at State DOTs

Generally, we found that DOTs provided much less information about their CMAQ processes than MPOs, which we attribute to the fact that the vast majority of CMAQ funds are programmed by MPOs. Many DOT websites provided very little, if any, information about their CMAQ programs. These were generally DOTs that received nearly all of their CMAQ apportionment as flexible funds and opted to roll their CMAQ funds into their STP and STIP processes. In other cases, DOTs provided very little information because they passed CMAQ funds along with most programming responsibility to MPOs. However, some DOTs do program CMAQ funds and in these cases information about project solicitation, rating, and selection processes were provided on their websites. The remainder of this section summarizes what we found for these DOTs, focusing on broad trends and unique approaches.

Project Solicitation Methods. Most DOTs issued annual or biannual calls for CMAQ projects. Calls often included application forms, information about project eligibility, and general information about the CMAQ program. Calls typically went out to MPOs, regional DOT offices, and municipalities.

Many DOT websites that we reviewed did not provide any information about how CMAQ projects are solicited. This was generally the case for DOTs that rolled the CMAQ evaluation and selection process into their overall STIP or STP project selection process, or passed CMAQ funds through to MPOs to program. For example, Nebraska did not have a formal CMAQ process. The Nebraska DOT states that it generally uses CMAQ funds for STP eligible projects; however, the DOT does accept unsolicited letters requesting CMAQ funding. In Kansas, the DOT devolved all CMAQ project selection tasks to MPOs in the state. DOTs may also lack a CMAQ project solicitation because the DOT withholds the funds for specific state projects. For example, the Rhode Island DOT used all of its CMAQ funds to support intercity commuter rail and bus transit.

From the set of DOTs that managed a project solicitation process, we identified several practices that we consider examples of good practice. Some DOTs made greater efforts to broadly solicit calls for proposals. For example, in New Hampshire, the DOT went as far as posting the call in the news media and holds pre-application workshops for potential project sponsors. The pre-application workshops provided an opportunity for potential project sponsors to ask questions about project eligibility, the application process, and what information they will need to provide. Several DOTs specifically reminded potential project sponsors that non-governmental organizations may also propose projects by partnering with a local or regional government agency. Additionally, many DOTs provided information on their websites or with their application packages, about how projects would be evaluated. In the simplest example, a list of funding priorities was provided as was the case in Montana. In other cases, detailed tables of project scoring and weighting schemes were provided, which we describe in more detail in the following section. Taken together, these practices are likely to result in a more diverse, yet responsive, pool of CMAQ project proposals as they aim to communicate clearly to all potential project sponsors what activities are eligible and the priorities of the DOT.

Most DOTs had specific CMAQ project application forms that were distributed with their calls for CMAQ proposals or that were available on their websites. For the most part, the forms are similar in that they ask for standard information about the project such as a broad description of the activities that will take place, where the project is located, sponsoring agencies, a budget, if the project is included in an TIP or STIP, and which CMAQ-eligible activity the project corresponds to. While each application also asks for information about the project's effect on congestion and air quality, the level of detail varies. The most streamlined applications, such as that used by Wisconsin, request a narrative explaining the project's impacts on congestion and air quality, but does not specifically require calculations or an explanation of calculation methods and assumptions. In other cases, such as in Kentucky and Iowa, the DOT specifically asks that calculations be provided along with assumptions. In still other cases, such as New Hampshire and Indiana, the DOT requires that local project sponsors work with MPOs or the DOT to complete their air quality analysis. Montana provides spreadsheet models and Michigan provides Adobe Acrobat forms that provide emission estimates for different project types. Users must still enter some data that require estimation and assumptions, such as the amount of VMT that is reduced; however, at a minimum the emission modeling steps are standardized.

We consider applications that request calculations for any quantitative information along with data sources and descriptions of all assumptions to be examples of the best practices observed in the DOTs we reviewed. Spreadsheet models or forms that produce calculations are also good tools for standardizing some aspects of congestion and emission reduction estimates, but they generally still require many data inputs that must also be estimated or modeled and that also require numerous assumptions. Spreadsheet models and similar tools also may not cover every possible project type. These tools alone do not necessarily replace the need for a clear explanation of calculation methods, data sources, and assumptions. The approach used by New Hampshire and Indiana may also be a good practice; however, that depends on whether more standardized and transparent, and potentially more accurate, estimates outweigh the cost of the greater burden placed on DOT staff, their consultants, and other resources.

Scoring and Ranking Schemes. Projects that have been proposed by local sponsors and MPOs are generally rated in some way and then ranked in order of funding priority. Two methods were common: scoring and weighting schemes and project priority rubrics or lists. Some DOTs had no formal process outside of the process used for general TIP or STIP project selection. These DOTs were generally those that received all CMAQ funds as flexible funds.

The most common method was scoring each project by assigning points based on how well the project responds to individual criteria. The size of emissions reductions and cost effectiveness in emissions reduction were important components of all scoring and ranking schemes. A project sponsor's record of completing projects, or completing them on time, was also a common criterion. However, other criteria that may be scored and their relative weight vary widely. States such as Wisconsin had a simple scoring scheme that focuses mostly on cost effectiveness. States at the other end of the spectrum, such as Indiana, had a long list of detailed scoring criteria that result in cost effectiveness accounting for only up to 20 points out a total possible score of 90.

An alternative approach to explicitly scoring proposals based on individual criteria is to develop a prioritized list of project types or a rubric that prioritizes projects in several different categories. This approach can be as simple as the ordered list of funding priorities used by the Montana DOT. It can also be a rubric, which prioritizes project types in several general categories. For example, Louisiana used a rubric that classifies projects into three levels of priority (high, medium, and low) for five categories of project types: traffic flow and intelligent transportation systems (ITS), alternative fuel/diesel retrofits, transit/diesel retrofits, TDM, and bike/pedestrian).

There are pros and cons to each approach. Systems that score projects based on set criteria can be more transparent and may seem more objective. Project sponsors know how their project will be evaluated and the relative importance of each component to the DOT. However, scoring systems can be inflexible. A rigid scoring system may not capture the benefits of a particular project, for example, a project that reduces emissions near a school or daycare center where air quality improvements would be more beneficial to these sensitive receptors. An additional concern with scoring schemes is that how the scheme is created can affect which projects are prioritized. For example, a congestion relief project with a low cost effectiveness would score very low using Wisconsin's scoring scheme and potentially very high using Indiana's. Therefore, it is important that the process of creating scoring schemes be transparent in addition to how the scores are used.

Prioritized lists and rubrics, like scoring schemes, let project sponsors know the priorities of the DOT but are generally more flexible. Methods that avoid assigning points, or minimize the number of project attributes or outcomes that are explicitly scored, leave more room for DOTs to weigh the relative importance of different factors on a case-by-case basis. The effectiveness of most CMAQ projects and the relative importance of other project outcomes is highly context sensitive. A project evaluation method that allows for the consideration of each project's unique context is therefore desirable. The main tradeoff between each approach is a greater sense of objectivity versus a potentially more robust and flexible, albeit potentially more subjective, process.

In either case, the project selection process ultimately depends on the DOT's final judgment; where the project ranks, whether determined by a score or otherwise, makes up only part of the overall decision. However, arguably, a score seems to carry greater weight and it may be hard to decline a highly-scored project that for other reasons is not as attractive as projects that were scored lower. The best practices, in our judgment, are those that have a scoring scheme or prioritization scheme that provides enough detail to communicate to sponsors the priorities of the agency, that maintains the focus on the CMAQ program's primary aim of emissions reductions and cost effectiveness, yet allows flexibility for the DOT to consider each project's specific context and unique features. A more flexible process is also less likely to discourage potentially unique and innovative projects.

Project Selection. Once projects have been prioritized in some manner as described above, a final decision must be made. A final decision typically considered prior scoring and ranking exercises but also considered other attributes of projects that may not have been considered. Typically, geographical equity was considered so that all funds do not flow to the same municipalities or regions. It was also common to ensure a variety of project types are selected (e.g., modal diversity). If the rating process did not consider a sponsor's past performance on completing projects, that was also a common consideration at this stage. The quality of proposals was also often evaluated; for example, how likely it is that the purported benefits will be realized based on the data and calculation methods that were used.

Two governance structures were most common in making these final vetting decisions: CMAQ-specific committees and general DOT boards that are also responsible for vetting other types of projects. In either case, a final decision was reserved for a state's transportation commission, commissioner, or equivalent high-level authority.

About half of the DOTs we reviewed that manage their CMAQ program used a CMAQ-specific project selection committee to accomplish this final project vetting. In some cases, the committee also screened pre-applications or even worked with project sponsors on their applications. CMAQ committees (which go by various names) included a broad range of members that typically include DOT engineering and planning staff, a representative from the state's environmental department, and a representative from FHWA. It was also common to include members from MPOs. For example, New Hampshire included representatives from each of the state's nine regional planning commissions. Two DOTs also included members representing transit interests (DOT transit engineers or transit agency officials). Other DOTs used more general project selection boards, rolling the vetting of CMAQ projects into their overall STIP or STP process. In other cases, such as Kansas, all decision making was devolved to MPOs.

The most important component of either a CMAQ-specific or more general DOT project selection committee is the makeup of the committee. The unique aspect of the CMAQ program is its focus on air quality and the requirement that emission reductions be estimated. Therefore, any committee should have members who are experts in vehicle emission modeling and mobile source air quality issues. Other aspects of CMAQ applications are not very different than other transportation project proposals and therefore there does not appear to be a strong argument for or against CMAQ-specific project panels.

CMAQ Project Selection Practices at Metropolitan Planning Organizations

We also evaluated how MPOs programmed their CMAQ funds. Since the focus of our study was on DOTs and there are many MPOs we only evaluated a small sample of them. We used three different approaches for selecting MPOs.

First, we evaluated four MPOs that received some amount of flexible CMAQ funds. They were the Puget Sound Regional Council (PSRC), Southeastern Wisconsin PRC, Metropolitan Council of the Twin Cities, and South Jersey Transportation Planning Organization. Note that relatively few MPOs actually received flexible funds based on our analysis of the FHWA CMAQ database. These MPOs were selected because they represent urban areas of different size and geographic location.

We also reviewed four MPOs that were likely to represent good or best practices based on our prior knowledge of their activities, evaluations made in prior studies, and the response from our inquiry to the Transportation and Air Quality committee of the Transportation Research Board. These were the Chicago Metropolitan Agency for Planning (CMAP), Delaware Valley Regional Planning Commission (DVRPC), Atlanta Regional Commission (ARC) and Wasatch Front Regional Council (WFRC).

Finally, we sent an e-mail request to a list-serve maintained by the Transportation and Air Quality committee of the Transportation Research Board requesting information on innovative and best practices in CMAQ programming. While this request did not specifically request information from DOTs or MPOs, most of the feedback concerned the practices of MPOs. The purpose of the e-mail request was to identify other unique or potentially robust practices that our literature review and scan of DOT and MPO websites may have missed. We used the results of this process to help select the MPOs above that were likely to represent best practices. We followed up with most of those who responded to our e-mail inquiry by phone or e-mail to better understand how they viewed their process and what they believed were specific strengths and weaknesses. The following section summarizes our findings.

Project Solicitation. Methods for soliciting CMAQ projects were generally the same as those used by DOTs. Many MPOs issue CMAQ-specific calls while others include them within their general TIP process. In some cases, DOTs issue the call but project sponsors submit proposals through MPOs. Calls went out to local agencies and were often publicly posted on the MPO's website. One notable process that we believe is an example of good practice is that used by ARC. ARC coordinated with state and local government officials to develop the emphasis areas for its CMAQ call. This was the only example in all the DOTs and MPOs we reviewed that mentioned a multi-agency process for developing a CMAQ call for proposals. Other DOTs and MPOs may have done something similar but did not provide any information.

MPOs were much more likely to have detailed application packages for CMAQ projects. Application packages typically included detailed instructions, emphasized priority project types, and included standardized forms for estimating emission and other project benefits. Many MPOs had forms or other tools that automatically estimated emission benefits and cost effectiveness. ARC had one of the most comprehensive and detailed emission estimation calculators, which estimated changes in VMT, hours of delay, criteria, and greenhouse gas emissions. ARC staff were also available to help local project sponsors collect the required input data. WFRC also used a detailed calculator that is very similar to ARC's. Similar to some DOTs, two of the MPOs we evaluated request information from project sponsors which was then used to estimate emission benefits and cost effectiveness. The PSRC, in particular, requested very detailed data from project sponsors which laid out data requirements for each project type so that PSRC staff could estimate emission reductions. MPOs in general also placed more emphasis on integration of projects with the region's long-range regional transportation plan.

Overall, MPOs appeared to have more comprehensive project solicitation processes than DOTs, particularly with regard to application materials and tools.

Scoring and Ranking Schemes. Almost all MPOs that we reviewed rated proposals using a system of scores and weights. In general, the scoring and weighting schemes were also more detailed and comprehensive than those used by the DOTs we evaluated. In some cases, the scoring and weighting schemes applied to all projects (not only CMAQ). In these cases, there were usually categories relevant to CMAQ projects, such as emission reductions and cost effectiveness of emission reduction. As mentioned above, many MPOs also provided detailed application instructions that also contained information about each MPO's priorities, which sometimes contained ranked lists and rubrics of priority project types.

Our review of MPOs revealed several examples of scoring methods that were not covered in our review of DOT methods. CMAP scored CMAQ proposals using categories and points that differ by project type. For example, transit projects were scored based on ridership (15 points) and reliability or asset condition (15 points), while bicycle projects were scored based on safety and attractiveness (10 points), transit access (10 points), and facility connectivity (10 points). CMAP's method tailors the scoring scheme to what is more relevant for different types of projects. ARC used a unique process for determining criteria weights. They surveyed local government staff and asked them to rank each criterion from most to least important. The result was a two-dimensional rubric that indicates the proportion of weight given to each of 12 criteria for each project type. This scoring scheme was used for all programs, not just CMAQ. ARC's process avoids some of the subjectivity and potential bias that can occur when creating scoring and weighting schemes. DVRPC was unique among the DOTs and MPOs we evaluated in that its scoring scheme was the only one that included points for addressing environmental justice, which is generally a major concern with regard to mobile source emissions (*7–9*).

Project Selection. Overall, the final vetting and selection of CMAQ projects was similar to the procedures used by the DOTs we evaluated.

Some MPOs used CMAQ-specific selection committees. These committees typically included transportation planners and engineers, transit agency staff, and in some cases air quality or environmental agency staff. They also often included representatives of local government and also FHWA and DOT staff. DVRPC was unique in that its CMAQ committee also included members representing business interests and the general public. The Southern Wisconsin RPC has all CMAQ projects reviewed independently by the planning commission's staff, Wisconsin DOT, and the Wisconsin Department of Natural Resources. These agencies then meet and agree on a final list.

In other cases, the MPO's usual process for evaluating TIP projects was used rather than a CMAQ-specific committee. For MPOs such as ARC, where air quality is a major transportation planning concern, the use of a more general project selection is not necessarily a limitation because air quality considerations are a normal part of the planning process. For MPOs with relatively few air quality concerns, a more general project selection process may fail to properly vet the emission reduction benefits of projects seeking CMAQ funds.

Finally, as was the case for DOTs, after a final round of vetting the ultimate decision for programming CMAQ projects was made by the MPO's board or the state DOT.

Conclusion

We were challenged by NMDOT to identify five DOTs and MPOs that represent the best overall examples of practice for programming CMAQ projects at the statewide level and recommending a general assessment framework. No single DOT or MPO in our opinion offers a good example of best practice for most DOTs. Some of the MPOs with the most robust practices face transportation and air quality challenges of both a very different scale and context than most DOTs. For example, ARC has one of the most comprehensive CMAQ programming processes out of the DOTs and MPOs that we reviewed. However, ARC's methods are tailored to the challenges of planning a large urban area with severe congestion and air quality problems. ARC staff have specific expertise in specific areas such as air quality and transit modeling and access to state of the art modeling tools. DOTs, particularly those receiving mostly flexible funds, often face a very different set of problems: a sparsely populated state with relatively minor congestion and air quality problems. Within this context, many DOTs and the communities they serve likely have fewer resources than ARC or other large metropolitan areas for evaluating the air quality impacts of such a wide variety of project types. While we identify five MPOs and DOTs that we believe have the best overall practices, we caution that not every practice is one we consider to be a good example and not every practice would be applicable to every DOT.

Best Practices

ARC has a very comprehensive process, both for CMAQ programming and for their overall TIP. What is particularly impressive is ARC's robust emission and congestion benefit calculator, their method for creating a scoring and weighting scheme based on the input of staff from municipalities in the region, and their multi-tiered project screening process that ensures projects fit within ARC's overall plan for the region. ARC does not use a CMAQ-specific selection committee, but provides a relatively large amount of CMAQ-specific information and tools for project sponsors to use. Air quality and congestion are two factors that are important in all ARC projects.

New York State DOT (NYSDOT) is selected for many of the same reasons as ARC. They have a very comprehensive CMAQ process including a detailed emission benefit calculator. NYSDOT also distributes a concise CMAQ guidebook to project sponsors that explains basic project eligibility and the NYSDOT application and evaluation process. Furthermore, NYSDOT requires that project sponsors attend a mandatory workshop that further explains the application process and expectations. One difference from ARC is that NYSDOT requests information from project sponsors which they then input into their emission benefit calculation tool. NYSDOT also withholds some CMAQ funds for statewide projects and programs. The efforts NYSDOT makes to inform potential project sponsors with concise guidance documents and workshops are practices that we think are important for all DOTs. We also think that the approach where NYSDOT performs emission calculations may be a more realistic approach for DOTs in states with many rural communities and small towns with fewer technical resources than the procedures used by many other DOTs and MPOs where individual project sponsors must make these estimates.

We include the Michigan DOT because it has the most comprehensive and tailored set of CMAQ project application forms for a DOT. The forms automatically perform some calculations, structure required inputs, and require no specialized software. The use of these detailed forms could help staff from small towns and rural communities perform detailed emission reduction calculations. However, the forms do require numerous inputs that may still represent a significant burden for some agencies.

We also include New Hampshire DOT (NHDOT) because of their significant outreach efforts. New Hampshire DOT distributes CMAQ requests for proposals widely; for example, by issuing press releases to major media outlets. This recruitment effort is followed by pre-application workshops that provided additional information about the CMAQ program, its requirements, how proposals will be evaluated and information that project sponsors will have to provide. Furthermore, NHDOT requests information from project sponsors which it then uses to perform emission reduction estimates, which reduces the burden on the many small and rural communities within the state. Additionally, NHDOT uses a CMAQ selection committee composed of representatives from every region in the state, which may ensure some amount of regional equity in the process and local expertise in specific regional priorities.

Finally, we include Louisiana as an example of the best alternative to the scoring and weighting schemes that were very common for both DOTs and MPOs. The DOT in Louisiana uses a priority rubric. Project types in five general categories are assigned one of three qualitative priorities. The rubric provides more flexibility to consider unique projects by not assigning predetermined weights to very specific criteria. Rating projects in different categories also seems more useful. For example, diesel retrofits have been shown to be very cost effective at reducing PM2.5 emissions; however, if PM2.5 is not a local issue or if there is not a lot of heavy-duty diesel truck traffic in the area, then this strategy's high effectiveness is not very relevant. Most DOTs and MPOs expressed a desire to fund projects of various types; a rubric is one way to help accomplish that goal.

Considering Exposure

Emission reductions from CMAQ projects may lower pollution exposure and reduce health risks. However, the link between emissions reductions and broader benefits depends on factors that are not routinely considered in CMAQ project evaluations. Criteria pollutants such as carbon monoxide and primary particle pollution form steep concentration gradients along roadways (*10*), limiting emission reduction benefits to a relatively narrow area. Other criteria air pollutants such as nitrogen oxides and volatile organic compounds, which are ozone precursors, contribute to regional air quality impacts. While a full-scale air quality modeling analysis may be out of reach for most project sponsors, the spatial extent of benefits could be considered qualitatively based on what is currently understood about each type of pollutant.

Author Contributions

The authors confirm contribution to the paper as follows: study conception and design: Gregory Rowangould; data collection: Amir Poorfakhraei, Razieh Nadafianshahamabadi; analysis and interpretation of results: Razieh Nadafianshahamabadi, Amir Poorfakhraei, Gregory Rowangould; draft manuscript preparation: Gregory Rowangould, Razieh Nadafianshahamabadi, Amir Poorfakhraei. All authors reviewed the results and approved the final version of the manuscript.

References

1. *Special Report 264: The Congestion Mitigation and Air Quality Improvement Program: Assessing 10 Years of Experience.* TRB, National Research Council, Washington, D.C., 2002.
2. FHWA. *SAFETEA-LU 1808: Congestion Mitigation and Air Quality Improvement Program Evaluation and Assessment: Phase II Final Report.* Publication FHWA-HEP-09-026. Prepared by USDOT Volpe National Transportation Systems Center for the Federal Highway Administration, Washington, D.C., 2009.
3. FHWA. *SAFETEA-LU 1808: Congestion Mitigation and Air Quality Improvement Program Evaluation and Assessment - Phase 1 Final Report.* Publication FHWA-HEP-08-019. Prepared by ICF International for the Federal Highway Administration, Washington, D.C., 2008.
4. FHWA. *Congestion Mitigation and Air Quality (CMAQ) Improvement Program Cost-Effectiveness Tables Development and Methodology.* Prepared by USDOT Volpe National Transportation Systems Center for the Federal Highway Administration, Washington, D.C., 2015.
5. Adler, K., M. Grant, and W. Schroeer. Emissions Reduction Potential of the Congestion Mitigation and Air Quality Improvement

Program: A Preliminary Assessment. *Transportation Research Record: Journal of the Transportation Research Board*, 1998. 1641: 81–88.
6. FHWA. The Congestion Mitigation and Air Quality (CMAQ) Improvement Program Under the Moving Ahead for Progress in the 21st Century Act: Interim Program Guidance. 2013. US DOT Federal Highway Administration.
7. Houston, D., W. Li, and J. Wu. Disparities in Exposure to Automobile and Truck Traffic and Vehicle Emissions near the Los Angeles–Long Beach Port Complex. *American Journal of Public Health*, Vol. 104, No. 1, 2014, pp. 156–164. https://doi.org/10.2105/AJPH.2012.301120
8. Rowangould, G. A Census of the US Near-Roadway Population: Public Health and Environmental Justice Considerations. *Transportation Research Part D: Transport and Environment*, Vol. 25, 2013, pp. 59–67.
9. Tian, N., J. Xue, and T. M. Barzyk. Evaluating Socioeconomic and Racial Differences in Traffic-Related Metrics in the United States Using a GIS Approach. *Journal of Exposure Science and Environmental Epidemiology*, Vol. 23, No. 2, 2013, pp. 215–222.
10. Karner, A. A., D. S. Eisinger, and D. A. Niemeier. Near-Roadway Air Quality: Synthesizing the Findings from Real-World Data. *Environmental Science & Technology*, Vol. 44, No. 14, 2010, pp. 5334–5344.

The Standing Committee on Transportation Programming and Investment Decision-Making (ADA50) peer-reviewed this paper (18-06697).

The views expressed in this paper are solely those of the authors and do not necessarily reflect those of NMDOT.

Sage Publications USA
sagepub.com

ISBN 978-1-5108-8256-0
9 781510 882560

CURRAN ASSOCIATES INC.
proceedings.com